Niels Hartanto

m-Monic Operator- and Matrix Polynomials

Niels Hartanto

m-Monic Operator- and Matrix Polynomials

Spectral Properties, Factorizations and Companion Forms

Südwestdeutscher Verlag für Hochschulschriften

Impressum/Imprint (nur für Deutschland/only for Germany)
Bibliografische Information der Deutschen Nationalbibliothek: Die Deutsche Nationalbibliothek verzeichnet diese Publikation in der Deutschen Nationalbibliografie; detaillierte bibliografische Daten sind im Internet über http://dnb.d-nb.de abrufbar.
Alle in diesem Buch genannten Marken und Produktnamen unterliegen warenzeichen-, marken- oder patentrechtlichem Schutz bzw. sind Warenzeichen oder eingetragene Warenzeichen der jeweiligen Inhaber. Die Wiedergabe von Marken, Produktnamen, Gebrauchsnamen, Handelsnamen, Warenbezeichnungen u.s.w. in diesem Werk berechtigt auch ohne besondere Kennzeichnung nicht zu der Annahme, dass solche Namen im Sinne der Warenzeichen- und Markenschutzgesetzgebung als frei zu betrachten wären und daher von jedermann benutzt werden dürften.

Verlag: Südwestdeutscher Verlag für Hochschulschriften GmbH & Co. KG
Dudweiler Landstr. 99, 66123 Saarbrücken, Deutschland
Telefon +49 681 37 20 271-1, Telefax +49 681 37 20 271-0
Email: info@svh-verlag.de

Approved by: Berlin, TU, Diss., 2011

Herstellung in Deutschland:
Schaltungsdienst Lange o.H.G., Berlin
Books on Demand GmbH, Norderstedt
Reha GmbH, Saarbrücken
Amazon Distribution GmbH, Leipzig
ISBN: 978-3-8381-2937-2

Imprint (only for USA, GB)
Bibliographic information published by the Deutsche Nationalbibliothek: The Deutsche Nationalbibliothek lists this publication in the Deutsche Nationalbibliografie; detailed bibliographic data are available in the Internet at http://dnb.d-nb.de.
Any brand names and product names mentioned in this book are subject to trademark, brand or patent protection and are trademarks or registered trademarks of their respective holders. The use of brand names, product names, common names, trade names, product descriptions etc. even without a particular marking in this works is in no way to be construed to mean that such names may be regarded as unrestricted in respect of trademark and brand protection legislation and could thus be used by anyone.

Publisher: Südwestdeutscher Verlag für Hochschulschriften GmbH & Co. KG
Dudweiler Landstr. 99, 66123 Saarbrücken, Germany
Phone +49 681 37 20 271-1, Fax +49 681 37 20 271-0
Email: info@svh-verlag.de

Printed in the U.S.A.
Printed in the U.K. by (see last page)
ISBN: 978-3-8381-2937-2

"Cruelty to animals is one of the most significant vices of a low and ignoble people. Wherever one notices them, they constitute a sure sign of ignorance and brutality which cannot be painted over even by all the evidence of wealth and luxury. Cruelty to animals cannot exist together with true education and true learning."
– Alexander von Humboldt

†

"I am not interested to know whether vivisection produces results that are profitable to the human race. The pain which it inflicts upon unconsenting animals is the basis of my enmity toward it, and it is to me sufficient justification of the enmity without looking further".
– Mark Twain

†

"Imagine living your life in a small, filthy cage constantly in pain, unable to stand or lie down comfortably. After months of agony, your torture finally ends, but not at the slaughterhouse. Instead, two gentle hands reach down to lift you out of the darkness, and bring you to a safe, loving place. For the first time in your life you can stretch your wings and legs and feel soft straw and cool grass beneath your feet."
– Dr. Karen Davis

†

Acknowledgements

Die Fertigstellung dieser Arbeit wäre nicht möglich gewesen ohne die Hilfe und Unterstützung einiger Menschen, die ich hier dankend erwähnen möchte.

Zunächst möchte ich mich ganz herzlich bei Karl-Heinz Förster bedanken. Das Thema der Dissertation, dessen Richtung er vorgegeben hat, war spannend und bereitete mir viel Freude. Besonders dankbar bin ich für die stetige und geduldige Unterstützung in Form von vielen langen Diskussionen, hilfreichen Anregungen und konstruktiver Kritik und für die vielen motivierenden Impulse.

Zu besonders großem Dank, nicht nur für die Begutachtung dieser Dissertation, bin ich Volker Mehrmann verpflichtet. Seine anfänglichen Vorschläge und Anregungen haben die spätere Richtung der Arbeit wesentlich beeinflusst. Die fachlichen Gespräche mit ihm waren stets enorm effizient und hilfreich und seine Kritik in Bezug auf das Verfassen von wissenschaftlichen Arbeiten sind für mich von sehr hohem Wert gewesen.

Sehr dankbar bin ich auch Christian Mehl und Michael Karow für die hilfreichen Gespräche und Diskussionen und Bela Nagy für die Begutachtung dieser Arbeit.

I would like to thank Dario Bini very much for active and invaluable email discussions and Federico Poloni for the inspiring exchange.

Meiner Familie möchte ich großen Dank für die private Unterstützung aussprechen. Ohne sie hätte diese Arbeit nicht entstehen können. Magda Nafalska danke ich sehr für das Korrekturlesen und für ihre stets offenen Ohren und motivierenden Worte in etwas schwierigeren Zeiten.

Contents

Introduction

Let $m \in \mathbb{N}$, $m \geqslant 1$ and $G \subset \mathbb{C}$ be a domain and let $\mathfrak{A}$ be a complex Banach algebra. We consider functions $F : G \to \mathfrak{A}$ of the type

$$F(\lambda) = \lambda^m I - A(\lambda), \tag{I}$$

where I denotes the identity in $\mathfrak{A}$ and $A(\lambda) = \sum_{j\in\mathbb{Z}} \lambda^j A_j$, $A_j \in \mathfrak{A}$, is a Laurent series. We will call functions of this type m–monic functions. In applications various special cases of functions of this type appear.

For instance, in modeling of queueing problems, where Markov chains play a major role, matrix functions of the type

$$F(\lambda) = \lambda I - \sum_{j=-1}^{\infty} \lambda^{j+1} A_j$$

with entrywise nonnegative $n \times n$ matrix coefficients A_j arise. They are associated with the transition matrix of the Markov chain the entries of which are the transition probabilities from one to each other state of the stochastic process. The coefficients satisfy the condition $(\sum_{j=-1}^{\infty} A_j)\mathbf{1}_n = \mathbf{1}_n$, where $\mathbf{1}_n$ denotes the vector in $\mathbb{R}^n$ the entries of which are all equal to 1. Very crucial for the analysis of these Markov chains is the minimal entrywise nonnegative solution of the matrix equation

$$X = \sum_{j=-1}^{\infty} A_j X^{j+1}.$$

It is strongly connected to certain factorizations of the function F and key to find so-called stationary vectors, which represent equilibrium probability distributions of the Markov chain. See e.g. [BLM05], [GHT96], [LR99], [Mei06], [Neu94], [Neu89].

Another field of application is in hydrodynamics. For instance, the study of small motions and normal oscillations of a viscous incompressible fluid in an open container leads to a spectral problem for functions of the type

$$L(\lambda) = I - \lambda A - \lambda^{-1} B,$$

where A is a positive definite and B a nonnegative definite operator. See e.g. [AKL68], [AHKM03], [KL68]. Multiplying with λ on both sides leads to the 1–monic function

$$F(\lambda) = \lambda L(\lambda) = \lambda I - (\lambda^2 A + B).$$

Another example can be found in [AKS97], [Suk97], where the investigation of small convective motions of a fluid in a container results in spectral problem for the operator function

$$L_\epsilon(\lambda) = \lambda^2 A - \lambda(\epsilon Q - I) + C = \lambda I - (-\lambda^2 A + \lambda\epsilon Q - C),$$

where A, Q, C are compact self adjoint operators in a Hilbert space, A is positive definite, C is nonnegative definite and ϵ is a positive parameter.

Another instance where m–monic functions arise implicitly is in [Mar88], where factorizations of functions with arbitrary operator coefficients are investigated.

However, m–monic functions have not been studied explicitly, yet. Clearly, for any arbitrarily chosen $m \in \mathbb{N}$, any function in Laurent representation can be rewritten as an m–monic function. Nevertheless, the m–monic form loses this arbitrariness when the coefficients of A are supposed to have special properties like entrywise nonnegativity, nonnegative definiteness.

The intention of this thesis is to study spectral properties of m–monic functions where the coefficients A_j are elements of a complex Banach algebra $\mathfrak{A}$. A major focus lies on the two cases where the coefficients are linear bounded self adjoint operators in a Hilbert space or entrywise nonnegative $n \times n$-matrices. Furthermore, we study very closely the case when F is a polynomial.

The behavior of the spectral radius of $A(\lambda)$ as λ varies through G has a considerable impact on the structure and distribution of the spectral points of F. We exploit this connection by investigating the real valued function

$$\phi_A\colon G \cap [0,\infty) \to \mathbb{R}^+, \quad \phi_A(\tau) = \max_{|\lambda|=\tau} \operatorname{spr} A(\lambda),$$

where $\operatorname{spr} A(\lambda)$ denotes the spectral radius of $A(\lambda)$.

This strategy has already been used for special cases in [FN05a] to study spectral properties of quadratic 1– monic matrix polynomials where the coefficients of A are entrywise nonnegative matrices and the sum of the coefficients is irreducible. Factorizations of m–monic polynomials are studied in [FN05b], where the coefficients are elements of a cone in an ordered Banach algebra.

A similar approach has been used by H. K. Wimmer and J. Swoboda. The peripheral eigenvalues of monic matrix polynomials (monic polynomials of degree m are m–monic) with Hermitian coefficients are investigated in [Wim08] under a condition which is closely related to the spectral radius of $|A|(\lambda)$, where $|A|(\lambda) := \sum_{j\in\mathbb{Z}} \lambda^j (A_j^2)^{1/2}$. [SW10] studies the spectrum of monic operator polynomials with bounded nonnegative definite coefficients under the same condition.

A special role in the study of spectral properties of m–monic functions is played by 1–monic functions. For certain problems, it is much easier to handle the 1–monic case using fixpoint iterations. Therefore, the subject of suitable transformations from m–monic polynomials to 1–monic ones is studied as well.

We will proceed as follows. In Section 1.1 we will consider arbitrary m–monic functions with coefficients in a Banach algebra, which are analytic on an annulus. We will present a general framework to study m–monic functions with coefficients that are either nonnegative definite linear bounded operators acting on a Hilbert space or entrywise nonnegative square matrices. The function ϕ_A is introduced in this section. It will be used to describe spectral properties of F, for instance to exclude certain regions from the spectrum of F, to give the distribution of spectral points with respect to certain circles, or to give a criterion for the existence of so-called spectral factorizations of F.
In Section 1.2 we will make some technical preparations for the investigation of matrix valued m–monic functions. We specify circles containing certain numbers of eigenvalues of F and give representations for the derivatives of the mapping $\det F : \tau \mapsto \det F(\tau)$ at the points where the function ϕ_A intersects with the mapping $\tau \mapsto \tau^m$. These results will be useful to study the eigenvalues of F on

circles with radii determined by these intersection points in later sections.
Section 1.3 is dedicated to the special case when F is a matrix polynomial. We give a short overview of eigenvalues and Jordan chains of matrix polynomials.

In Sections 2.1 and 2.2 we extend results of H. K. Wimmer and J. Swoboda, given in [Wim08] and [SW10]. It is shown there that monic polynomials $P(\lambda) = \lambda^m I - \sum_{j=0}^{m-1} \lambda^j A_j$ with self adjoint matrix coefficients or with bounded nonnegative definite operator coefficients which in both cases satisfy the condition $\sum_{j=0}^{m-1} |A_j| \leqslant I$, where $|A_j| = (A_j^* A_j)^{1/2} = (A_j^2)^{1/2}$, have eigenvalues on the unit circle that are rotation invariant with respect to angles corresponding to certain roots of unity.
We extend these results to m–monic operator functions that are analytic on an annulus with self adjoint coefficients. We impose a slightly different condition on the coefficients, more precisely, we suppose that there exists a positive real ρ such that

$$\sum_{j\in\mathbb{Z}} \rho^j |A_j| \leqslant \rho^m I. \tag{0-1}$$

For $\rho = 1$, this is the condition of H. K. Wimmer and J. Swoboda. We first prove that if $\lambda \in \mathbb{C}$ is an eigenvalue of F such that $|\lambda|$ satisfies the condition (0-1), then $|\lambda|$ is an eigenvalue of the function $F_{|A|}$, defined via $F_{|A|}(\lambda) = \lambda^m I - \sum_{j\in\mathbb{Z}} \lambda^j |A_j|$. If an additional condition concerning the connection of the coefficients A_j and $|A_j|$ is satisfied, then the reverse implication is also true. Further results deal with the rotation invariance of the eigenvalues of F the modulus of which satisfy condition (0-1). The corresponding angle of invariance is the greatest common divisor of all $m - j$ such that $A_{m-j}v \neq 0$, where v is a corresponding eigenvector of F.

In the treatment of nonlinear eigenvalue problems, degree reduction and, in particular, linearizations are a well proven tool to give access to their analysis. Since we also focus on matrix and operator polynomials, it might not be surprising that we bring into play degree reductions as well. Chapter 3 presents a special reduction of degree which greatly helps to deal with m–monic polynomials.
In Section 3.1 we consider general $\mathfrak{A}$–valued polynomials Q, where $\mathfrak{A}$ denotes an algebra, and give a degree reduction $\mathcal{Q}$ of Q which is suited to reduce m–monic polynomials to 1-monic ones. We will use this reduction in Chapter 4, when we deal with matrix polynomials with nonnegative entrywise coefficients. The considered reduction is a generalization of the well known companion form used for linearizations of matrix polynomials, see e.g. [GLR82], [Mar88], [Rod89] and its coefficients are elements in the algebra $\mathfrak{A}^{m,m}$ for some $m \in \mathbb{N}$. Similar to linearizations via the companion form and other linearizations (see e.g. [AV04], [MMMM06]), the reduction given in Section 3.1 is obtained via an equivalence transformation

$$Q(\lambda) \oplus \mathcal{I}_{m-1} = \mathcal{E}(\lambda)\mathcal{Q}(\lambda)\mathcal{F}(\lambda),$$

where $\mathcal{E}$ and $\mathcal{F}$ are unimodular $\mathfrak{A}^{m,m}$-valued polynomials, i. e. they are invertible for all $\lambda \in \mathbb{C}$. Reducing the degree of Q in the given way leaves its spectrum unchanged.
In Section 3.3 we will study the effect of the degree reduction on the Jordan structure of Q if the coefficients are square matrices. We will give formulas to calculate the Jordan chains of the reduction from the original polynomial and vice versa. Furthermore, we will see that while linearizing a polynomial via a unimodular transformation does not change the number of eigenvalues (finite and infinite, with multiplicities) of the original polynomial, this number will in general increase

for degree reductions via unimodular transformations to a degree greater than one. The objective in Section 3.2 is to establish a connection between factorizations

$$Q(\lambda) = \left(\sum_{j=0}^{k-1} \lambda^j B_j\right)\left(\lambda^{l-k+1} I - \sum_{j=0}^{l-k} \lambda^j C_j\right), \quad B_j, C_j \in \mathfrak{A}$$

of the polynomial Q and factorizations

$$\mathcal{Q}(\lambda) = \mathcal{B}(\lambda)(\lambda \mathcal{I}_m - \mathcal{C}), \quad \mathcal{B}(\lambda), \mathcal{C} \in \mathfrak{A}^{m,m} \tag{0-2}$$

of the reduction $\mathcal{Q}$. Here, l and k denote the degree of Q and $\mathcal{Q}$, respectively. We will see that if $\mathcal{Q}$ has a factorization of the type (0-2), then $\mathcal{C}$ has to be the companion matrix of the right factor. Furthermore, each of these factorizations can be calculated from the other one.

In Chapter 4 we will concentrate on matrix polynomials with entrywise nonnegative coefficients.

It is well known from Perron-Frobenius theory (see e.g. [HJ85],[Min88], [BP94]) that the peripheral spectrum of an entrywise nonnegative irreducible matrix A is rotation invariant with respect to certain angles. These angles correspond to the dth roots of unity, where d is the so-called index of imprimitivity of A.
In [FN05a], K. H. Förster and B. Nagy investigated quadratic $n \times n$–matrix polynomials $P(\lambda) = \lambda I - (\lambda^2 A_2 + \lambda A_1 + A_0)$ with entrywise nonnegative coefficients such that the sum of the coefficients is irreducible. They proved that eigenvalues which can be seen as "peripheral" in the sense that they are located on the boundary circles of a certain spectrum-free annulus, are rotation invariant. The corresponding angles of invariance are characterized via graph theoretical concepts and are associated with a generalization of the index of imprimitivity of a nonnegative matrix. We will extend this result to m–monic matrix polynomials of degree $l \in \mathbb{N}$, $l > m$. Analogously to [FN05a], this is done via convenient spectral factorizations

$$P(\lambda) = \left(I_n - \sum_{j=1}^{l-m} \lambda^j B_j\right) B_0 \left(\lambda^m I_n - \sum_{j=0}^{m-1} \lambda^j C_j\right) \tag{0-3}$$

of P, such that the coefficients B_j, C_j are also entrywise nonnegative and B_0 is a nonsingular M–matrix. We will call the right factor of this factorization a Perron-Frobenius factor of P.
For these factorizations it is crucial to study the case of 1-monic matrix polynomials, which we will do in Section 4.1. We will see that the existence of a Perron-Frobenius factor in the case $m = 1$ is strongly connected to the convergence of a fixpoint iteration.
In Section 4.2, the results from Chapter 3 and Section 4.1 are put together to prove that there exists a factorization of the type (0-3) of an m–monic matrix polynomial with entrywise nonnegative coefficients if and only if there exists a $\rho > 0$ such that $\operatorname{spr} A(\rho) < \rho^m$. This result was already proved by K. H. Förster and B. Nagy in [FN05b] in the more general setting where the coefficients are elements of a closed normal algebra cone in an ordered Banach algebra. The proof relies on an abstract factorization result in ordered decomposing Banach algebras [GKS03]. We give a proof for the matrix case which completely relies on matrix theoretical concepts.
In Section 4.3 we associate with P an infinite graph and an integer d, the index of

phase imprimitivity, which determines the angles of invariance of the "peripheral" eigenvalues. Furthermore, we will see that with respect to the existence of factorizations of P of the type (0-3) and separation of eigenvalues of P, exactly one of eight possible cases holds, which we will also characterize in this section.

Chapter 5 presents a numerical method to calculate the factorization (0-3) of P, assumed it exists. It is mainly based on the method of cyclic reduction for certain Markov chains, as given by D. A. Bini, G. Latouche and B. Meini in [BLM05]. It is possible to apply this method to our setting if we transform the m–monic matrix polynomial P to a generating function of a convenient Markov chain. We will do this in Section 5.1. In Section 5.2, we will give the complete algorithm which calculates the coefficients of the desired factorization from the nonnegative coefficients of the given m–monic matrix polynomial.

In what follows we give frequently used notations for reference. Let T be an element of a Banach algebra $\mathfrak{A}$ and let E be a linear operator in a Hilbert space $\mathcal{H}$. Furthermore, let $A = (a_{ij})_{ij} \in \mathbb{C}^{n,n}$, $B = (b_{ij})_{ij} \in \mathbb{C}^{m,n}$, $D = (d_{ij})_{ij} \in \mathbb{C}^{m,n}$, $C = (c_{ij})_{ij} \in \mathbb{C}^{r,s}$.

$B(\mathcal{H})$	the space of bounded linear operators in $\mathcal{H}$
$\mathbb{D}$	the open unit disc $\{z \in \mathbb{C} \colon \lvert z\rvert < 1\}$
$\mathbb{D}_r$	the open disc $\{z \in \mathbb{C} \colon \lvert z\rvert < r\}$
$\mathbb{A}_{r,R}$	the open annulus $\{z \in \mathbb{C} \colon r < \lvert z\rvert < R\}$
$\mathbb{T}$	the unit circle $\{z \in \mathbb{C} \colon \lvert z\rvert = 1\}$
$\mathbb{T}_r$	the circle $\{z \in \mathbb{C} \colon \lvert z\rvert = r\}$
$\mathbb{E}_k$	the set $\{z \in \mathbb{C} : z^k = 1\}$ of the kth roots of unity
E^*	the adjoint of E w.r.t. to the inner product in $\mathcal{H}$
$\mathsf{N}(E)$	the kernel of E
$\mathsf{R}(E)$	the range of E
$\rho(T)$	the resolvent set of T
$\sigma(T)$	the spectrum of T
$\operatorname{spr} T$	the spectral radius of T
$\lvert A\rvert$	the matrix $(\lvert a_{ij}\rvert)_{ij}$, where $A = (a_{ij})_{ij}$
$\lvert E\rvert$	the operator $(EE^*)^{1/2}$
$\langle n\rangle$	the set $\{1,\dots,n\}$
$\langle n\rangle_0$	the set $\{0\} \cup \langle n\rangle$
$B \leqslant D$	$b_{ij} \leqslant d_{ij}$ for all $i \in \langle m\rangle$, $j \in \langle n\rangle$
$B < D$	$b_{ij} < d_{ij}$ for all $i \in \langle m\rangle$, $j \in \langle n\rangle$
$B \oplus C$	the block diagonal matrix $\left[\begin{array}{c\|c} A & 0 \\ \hline 0 & B \end{array}\right]$
$B \otimes C$	the Kronecker product $(b_{ij}C)_{ij} \in \mathbb{C}^{mr,ns}$ (for properties see e.g. [Bar83], [HJ91].)
$\mathbf{1}_n$	the vector $[1 \ \cdots \ 1]^T \in \mathbb{R}^n$
$\mathbf{1}_{n,n}$	the matrix $\mathbf{1}_n \otimes \mathbf{1}_n^T \in \mathbb{R}^{n,n}$

1 Spectral properties of algebra valued functions that are analytic in an annulus

1.1 Introduction and general results

In this first section, we give a short introduction to the fundamental notions and concepts of spectral theory in Banach algebras which we use throughout the whole thesis. Apart from that we assume the notation and basic properties of Banach algebras as given in [Con85].

$\mathfrak{A}$ denotes a complex Banach algebra with unit I, zero element 0 and the norm $\|\cdot\|$.

W.l.o.g. we may assume that $\|I\| = 1$, since by [Żel73, Corollary 2.5], there exists a norm $\|\cdot\|_I$ equivalent to the original norm $\|\cdot\|$ with $\|TS\|_I \leqslant \|T\|_I\|S\|_I$ and $\|I\|_I = 1$.

For an element T of $\mathfrak{A}$, the **spectrum** of T is is defined as the set

$$\sigma(T) = \{\lambda \in \mathbb{C}\colon \lambda I - T \text{ is not invertible in } \mathfrak{A}\}.$$

The spectrum $\sigma(T)$ of $T \in \mathfrak{A}$ is a compact set and, since we only consider complex Banach algebras, the spectrum of an element T of $\mathfrak{A}$ is never empty (see e.g. [Con85], [Aup91]). Furthermore, define the **spectral radius** $\operatorname{spr}(T)$ of T as

$$\operatorname{spr}(T) = \max\{|\lambda|\colon \lambda \in \sigma(T)\}.$$

Remember that $\operatorname{spr}(T) = \lim_{k\to\infty} \|T^k\|^{1/k}$ (see e.g. [Con85], [Aup91]).

For $0 < \eta \leqslant \infty$ denote by $\mathbb{D}_\eta$ the open disc

$$\mathbb{D}_\eta = \{\lambda \in \mathbb{C}\colon |\lambda| < \eta\},$$

where we set $\mathbb{D}_\infty = \mathbb{C}$. For $0 \leqslant \eta_1 < \eta_2 \leqslant \infty$ let

$$\mathbb{A}_{\eta_1,\eta_2} = \{\lambda \in \mathbb{C}\colon 0 \leqslant \eta_1 < |\lambda| < \eta_2\}$$

be the annulus with the inner radius η_1 and the outer radius η_2.

Consider a function $F : \mathbb{G} \to \mathfrak{A}$ which is analytic in $\mathbb{G}$, where $\mathbb{G}$ denotes either $\mathbb{A}_{\eta_1,\eta_2}$, $\mathbb{D}_\eta$ or $\mathbb{C}$. F has a Laurent series representation in $\mathbb{G}$, i. e.

$$F(\lambda) = \sum_{j\in\mathbb{Z}} \lambda^j A_j, \ \lambda \in \mathbb{G}, \tag{1-4}$$

with coefficients $A_j \in \mathfrak{A}$ for $j \in \mathbb{Z}$. As usual, we refer to $\sum_{j=1}^{\infty} \lambda^{-j} A_{-j}$ as the **principal part** of F and to $\sum_{j=0}^{\infty} \lambda^j A_j$ as the **regular part** of F.

Note that if $\mathbb{G} = \mathbb{C}$, then we have $A_j = 0$ for $j < 0$, i. e., F has a trivial principal part.

Define the **spectrum** $\sigma(F)$ of F as the set

$$\sigma(F) = \{\lambda \in \mathbb{C}\colon F(\lambda) \text{ is not invertible in } \mathfrak{A}\}.$$

Note that if $F(\lambda) = \lambda I - A$ with some $A \in \mathfrak{A}$, then $\sigma(F) = \sigma(A)$. Analogously to the spectral radius of elements of $\mathfrak{A}$ defined above, denote by

$$\operatorname{spr}(F) = \sup\{|\lambda|\colon \lambda \in \sigma(F)\}$$

the **spectral radius** of F

If, F is a function mapping from $\mathbb{G}$ into the space $B(\mathcal{H})$ of bounded linear operators in a Hilbert space $\mathcal{H}$, then a complex number $\lambda \in \sigma(F)$ is called an **eigenvalue** of F if there is a nonzero element $v \in \mathcal{H}$ such that $F(\lambda)v = 0$. Then v is called a corresponding **eigenvector**.

We call a real valued function f defined on any topological space D **upper semi-continuous** on D if the set $\{x \in D\colon f(x) < a\}$ is open for all $a \in \mathbb{R} \cup \{\pm\infty\}$. Let $G \subset \mathbb{C}$ be a region. A function $f : G \to \mathbb{R} \cup \{\infty\}$ is said to be **subharmonic** on G if it is upper semi-continuous on G and if it satisfies

$$f(\lambda_0) \leqslant \frac{1}{2\pi}\int_0^{2\pi} f(\lambda_0 + re^{i\varphi})d\varphi$$

for all $\lambda_0 \in G$ and for all $r > 0$ such that the closed disk $\overline{\mathbb{D}}_{\lambda_0,r} = \{\lambda \in \mathbb{C}\colon |\lambda - \lambda_0| \leqslant r\}$ is contained in G.

By a result of E. Vesentini (see [Aup91, Theorem 3.4.7]) the function $\lambda \mapsto \ln\left[\operatorname{spr}(A(\lambda))\right]$ is subharmonic on $\mathbb{A}_{\eta_1,\eta_2}$. From the theory of subharmonic functions (see [HK76, Theorem 2.13]) it then follows that the function

$$\tau \longmapsto \sup_{|\lambda|=\tau}\left(\ln\left[\operatorname{spr} A(\lambda)\right]\right) = \ln\left[\sup_{|\lambda|=\tau} \operatorname{spr} A(\lambda)\right], \quad \tau \in [0,\infty) \cap \mathbb{G}$$

is convex in $\ln\tau$ on (η_1, η_2). In other words, the function $t \longmapsto \ln\left[\max_{|\lambda|=e^t} \operatorname{spr} A(e^t)\right]$, $e^t \in [0,\infty) \cap \mathbb{G}$ is convex in t.

If we introduce the real valued function

$$\phi_A : \mathbb{G} \cap [0,\infty) \to \mathbb{R}^+, \quad \tau \longmapsto \sup_{|\lambda|=\tau} \operatorname{spr} A(\lambda),$$

then it follows that for $\tau_1, \tau_2 \in (\eta_1, \eta_2)$ and $\nu \in [0,1]$ the functional inequality

$$\phi_A({\tau_1}^{\nu}{\tau_2}^{1-\nu}) \leqslant (\phi_A(\tau_1))^{\nu}(\phi_A(\tau_2))^{1-\nu}$$

holds. Such functions are called **geometrically convex**, see e. g. [Kuc85]. The geometric convexity of ϕ_A and the functional inequality are rather crucial for us.

Recall that for a compact set $K \subset \mathbb{C}$, the **polynomially convex hull** $\hat{K}$ is defined to be set

$$\hat{K} = \{z \in \mathbb{C}\colon |p(z)| \leqslant \max_{u\in K}|p(u)| \text{ for all polynomials } p\}.$$

$\hat{K}$ is the union of K with the bounded components of $\mathbb{C} \setminus K$ or in other words $\hat{K}$ is obtained by filling any "holes" in K, see e.g. [Aup91], [Con85, Proposition 5.3]. K is said to be **polynomially convex** if $K = \hat{K}$. For more simple notations, in some occasions it is more convenient to write $K\hat{}$ instead of $\hat{K}$, e.g. $\sigma(A)\hat{}$, which we will do when appropriate.

Let T be an element of a Banach algebra $\mathfrak{A}$. The spectrum $\sigma(T)$ of T is contained in the closed disc $\overline{\mathbb{D}}_{\operatorname{spr}(T)}$. By definition of the polynomially convex hull and by considering the polynomial $p : z \mapsto z$, $z \in \mathbb{C}$, it follows that $\sigma(T)\hat{}$ does not contain any points outside of $\overline{\mathbb{D}}_{\operatorname{spr}(T)}$. This justifies the following remark.

Remark 1.1. *Let* $T \in \mathfrak{A}$. *Then* $\operatorname{spr} T = \sup\{|\lambda|\colon \lambda \in \sigma(T)\hat{}\}$.

Recall Liouville's Spectral Theorem, see for instance [Aup91, Theorem 3.4.14].

Theorem 1.2 (Liouville's Spectral Theorem). *Let A be an analytic function from $\mathbb{C}$ into a Banach algebra $\mathfrak{A}$. Suppose there exists a bounded set $M \subset \mathbb{C}$ such that $\sigma(A(\lambda)) \subset M$ for all $\lambda \in \mathbb{C}$. Then $\widehat{\sigma(A(\lambda))}$ is constant, i. e. it does not depend on λ.*

The following proposition follows from Theorem 1.2.

Proposition 1.3. *Let $\mathbb{G} = \mathbb{C}$ and consider A as in* (1-4)*. Then the following are equivalent.*

(*a*) *ϕ_A is bounded.*

(*b*) *ϕ_A is constant.*

(*c*) *$\sigma(A(\lambda))\hat{}$ is independent of λ, i. e. $\sigma(A(\lambda)) = \sigma(A_0)$.*

Proof. Suppose that (a) holds. Hence, there exists a $c \geqslant 0$ such that $\phi_A(\tau) \leqslant c$ for all $\tau \geqslant 0$. Due to $\operatorname{spr} A(\lambda) \leqslant \phi_A(|\lambda|) \leqslant c$ for all $\lambda \in \mathbb{C}$, we have $\sigma(A(\lambda)) \subset M$ for some bounded set $M \subset \mathbb{C}$. Theorem 1.2 then implies (c).

The implication $(c) \Rightarrow (b)$ follows from Remark 1.1 and $(b) \Rightarrow (a)$ is trivial. □

One important property of geometrically convex functions is the following.

Proposition 1.4. *Let A be as in* (1-4) *and suppose that there exist $\rho_1, \rho_2 \in \mathbb{R} \cap \mathbb{G}$, $0 < \rho_1 \leqslant \rho_2$ such that $\phi_A(\rho_j) = \rho_j^m$ for $j = 1, 2$. Then exactly one of the following assertions holds.*

(*i*) *$\phi_A(\rho) = \rho^m$ for all $\rho \in [\rho_1, \rho_2]$.*

(*ii*) *$\phi_A(\rho) < \rho^m$ for all $\rho \in (\rho_1, \rho_2)$.*

Proof. Since ϕ_A is geometrically convex, the function $\theta : \mathbb{R} \to \mathbb{R}$ with

$$\theta(\tau) = (\ln \circ \phi_A \circ \exp)(\tau) = \ln \phi_A(e^\tau)$$

is convex. Setting $\tau_j = \ln \rho_j$, $j = 1, 2$ the assumption $\phi_A(\rho_j) = \rho_j^m$ reads $\theta(\tau_j) = m\tau_j$. Due to the monotonicity of the exponential function, $\rho = e^\tau \in (\rho_1, \rho_2)$ if and only if $\tau \in (\tau_1, \tau_2)$. Hence, due to the convexity of θ, either

(*i*) $\theta(\tau) = m\tau$ for all $\tau \in [\tau_1, \tau_2]$, i.e. $\phi_A(\rho) = \rho^m$ for all $\rho \in [\rho_1, \rho_2]$

or

(*ii*) $\theta(\tau) < m\tau$ for all $\tau \in (\tau_1, \tau_2)$, i.e. $\phi_A(\rho) < \rho^m$ for all $\rho \in (\rho_1, \rho_2)$.

□

The following examples show that in Proposition 1.4 the assumption $\rho_1 > 0$ cannot be omitted. They can be found in [FN05a].

Example 1.5. Let $\mathfrak{A} = \mathbb{C}^{2,2}$, $p > 0$ and

$$A(\lambda) = \begin{bmatrix} \lambda^2 & 1 \\ \lambda p & \lambda^2 \end{bmatrix}.$$

For $\tau \geqslant 0$ we have $\operatorname{spr} A(\tau) = \tau^2 + \sqrt{\tau p}$. Setting $\tau = \xi^2$ for $\xi > 0$ we see that $\operatorname{spr} A(\tau) = \tau$ is equivalent to $\xi - \xi^3 = \sqrt{p}$. The function $[0, \infty) \to \mathbb{R}$, $\xi \mapsto \xi - \xi^3$ has its global maximum at $\xi_0 = \frac{1}{\sqrt{3}}$. Hence, the equation $\phi_A(\tau) = \tau$ or $\xi - \xi^3 = \sqrt{p}$ has exactly two positive solutions if and only if $\sqrt{p} < \xi_0 - \xi_0^3$ or $p < \frac{4}{27}$. In addition, $\tau = 0$ is always a solution, hence, for $p < \frac{4}{27}$, the equation $\operatorname{spr} A(\tau) = \tau$ has exactly three nonnegative solutions.

Example 1.6. Let $n \in \mathbb{N}$, $\mathfrak{A} = \mathbb{C}^{n,n}$, and $k_j \in \langle l \rangle_0 := \{0, 1, \ldots, l\}$ for $j \in \langle n \rangle$. For $\lambda \in \mathbb{C}$ consider the $n \times n$ matrix

$$A(\lambda) = \begin{bmatrix} 0 & \lambda^{k_1} & 0 & \cdots & 0 \\ \vdots & \ddots & \ddots & \ddots & \vdots \\ \vdots & & \ddots & \ddots & 0 \\ 0 & 0 & & \ddots & \lambda^{k_{n-1}} \\ \lambda^{k_n} & 0 & \cdots & \cdots & 0 \end{bmatrix}.$$

A is an $n \times n$ matrix polynomial with entrywise nonnegative coefficients. For $\lambda \neq 0$ the set of eigenvalues of the matrix $A(\lambda)$ is $\sigma(A(\lambda)) = \{z \in \mathbb{C} \colon z^n = \lambda^k\}$, where $k = \sum_{j=1}^n k_j$. Since the eigenvalues are all pairwise distinct, their algebraic and geometric multiplicities are 1. The corresponding eigenspaces of $A(\lambda)$ are $\operatorname{span}(\begin{bmatrix} 1 & \lambda^{-k_1} z & \lambda^{-k_1 - k_2} z^2 & \cdots & \lambda^{k_n - k} z^{n-1} \end{bmatrix}^T)$.
We have $\phi_A(\rho) = \operatorname{spr} A(\rho) = \rho^{\frac{k}{n}}$. Hence, for $k < n$, ϕ_A is concave and $\phi_A(\rho_1 = 0) = \rho_1$, $\phi_A(\rho_2 = 1) = \rho_2$.

We come now to the objects which are subject of our main interest. Fix $m \in \mathbb{Z}$, and define the $\mathfrak{A}$-valued function $F : \mathbb{G} \to \mathfrak{A}$ by

$$F(\lambda) = \lambda^m I - A(\lambda) = \lambda^m I - \sum_{j \in \mathbb{Z}} \lambda^j A_j, \tag{1-5}$$

which is also analytic in $\mathbb{G}$. Functions of this form play a major role in this thesis. We call such an F an m**–monic** function with coefficients A_j.

By means of the function ϕ_A we can easily exclude a certain region from the set of possible spectral points of the $\mathfrak{A}$-valued function F. This is the statement of the next Proposition.

Proposition 1.7. *Let F be as in* (1-5) *and let $\rho_1, \rho_2 \in [0, \infty) \cap \mathbb{G}$, $\rho_1 < \rho_2$, such that $\phi_A(\tau) < \tau^m$ for all $\tau \in (\rho_1, \rho_1)$. Then*

$$\sigma(F) \cap \mathbb{A}_{\rho_1, \rho_2} = \varnothing.$$

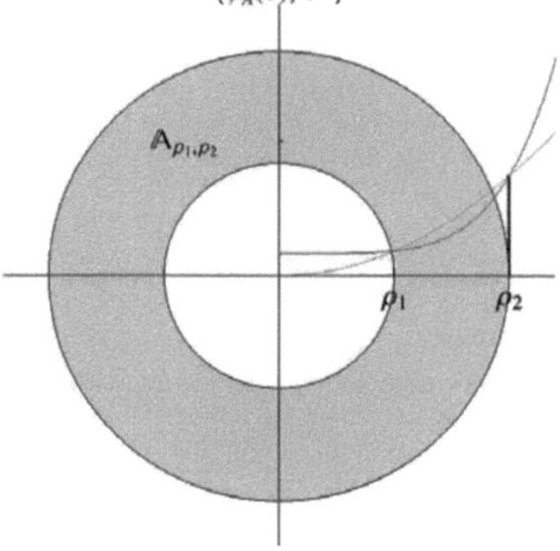

Figure 1: spectrum-free annulus $\mathbb{A}_{\rho_1,\rho_2}$

Proof. Suppose that there exists a $\mu \in \sigma(F) \cap \mathbb{A}_{\rho_1,\rho_2}$. This means that $|\mu| \in (\rho_1, \rho_2)$ and $\mu^m \in \sigma\big(A(\mu)\big)$, since $F(\lambda) = \lambda^m I - A(\lambda)$. Hence

$$|\mu|^m \leqslant \operatorname{spr} A(\mu) \leqslant \sup_{|\lambda|=|\mu|} \operatorname{spr} A(\lambda) = \phi_A(|\mu|) < |\mu|^m,$$

which is a contradiction. □

In the following we will establish conditions on the coefficients A_j under which ϕ_A satisfies

$$\phi_A(\tau) = \operatorname{spr} A(\tau) \in \sigma(A(\tau))$$

for all $\tau > 0$. This property is very important for our further analysis. For this purpose we will introduce cones in general Banach algebras and normal cones in ordered Banach algebras.

Denote by $\mathfrak{S}(\mathfrak{A}) = \{T \in \mathfrak{A}\colon \|T\| = 1\}$ the unit sphere in the Banach algebra $\mathfrak{A}$ and for $X \in \mathfrak{S}(\mathfrak{A})$ define

$$\mathfrak{S}_X(\mathfrak{A}') = \{f \in \mathfrak{A}'\colon f(X) = \|f\| = 1\},$$

where $\mathfrak{A}'$ denotes the dual space of $\mathfrak{A}$.

Note that due to the Hahn-Banach theorem, which states that each bounded linear functional defined on a subspace of a Banach space can be extended to the whole space leaving the norm unchanged (see e.g. [Con85], [Kat76]), $\mathfrak{S}_X(\mathfrak{A}')$ is nonempty for each $X \in \mathfrak{S}(\mathfrak{A})$. For $T \in \mathfrak{A}$ define the **numerical range** $\mathbb{V}(T)$ of T by

$$\mathbb{V}(T) = \{f(TX)\colon X \in \mathfrak{S}(\mathfrak{A}), f \in \mathfrak{S}_X(\mathfrak{A}')\}.$$

By [BD71, Lemma 2, Section 2], we have $\overline{\operatorname{conv}}\,\mathbb{V}(T) = \{f(T)\colon f \in \mathfrak{S}_I(\mathfrak{A}')\}$, where $\overline{\operatorname{conv}}\,\mathbb{V}(T)$ denotes the closure of the convex hull of $\mathbb{V}(T)$, i. e. the closure of the intersection of all convex sets containing $\mathbb{V}(T)$. We define the **numerical radius** $\operatorname{nr}(T)$ of T by

$$\operatorname{nr}(T) := \sup\{|\lambda|\colon \lambda \in \mathbb{V}(T)\} = \sup\{|f(T)|\colon f \in \mathfrak{S}_I(\mathfrak{A}')\}.$$

Note that, since by [BD71, Theorem 1, Section 10] we have $\sigma(T) \subset \overline{\mathbb{V}(T)}$, this implies that $\operatorname{spr}(T) \leqslant \operatorname{nr}(T)$. According to [BD71], [BD73], we call $T \in \mathfrak{A}$ **Hermitian** if $\mathbb{V}(T) \subset \mathbb{R}$. Furthermore, introduce the set

$$\mathfrak{P} = \{T \in \mathfrak{A} \colon \mathbb{V}(T) \subset \mathbb{R}^+\}.$$

We call $T \in \mathfrak{P}$ **nonnegative definite**. By definition, for $T \in \mathfrak{P}$ and $f \in \mathfrak{S}_I(\mathfrak{A}')$ we have that $f(T) \geqslant 0$. It is clear that a nonnegative definite element of $\mathfrak{A}$ is Hermitian.

A subset $\mathfrak{C} \subset \mathfrak{A}$ is called a **cone** if it satisfies

$$(i) \quad \mathfrak{C} + \mathfrak{C} \subset \mathfrak{C} \quad \text{and} \qquad (ii) \quad \lambda\mathfrak{C} \subset \mathfrak{C} \quad \text{for all } \lambda \geqslant 0 \qquad (iii) \quad I \in \mathfrak{C}.$$

If $\mathfrak{C}$ satisfies $-\mathfrak{C} \cap \mathfrak{C} = \{0\}$, then $\mathfrak{C}$ is called a **proper cone**. Obviously, $\mathfrak{P}$ is a cone in $\mathfrak{A}$. Furthermore, if we suppose that $T \in -\mathfrak{P} \cap \mathfrak{P}$, then this implies that $\operatorname{nr}(T) = 0$. By Proposition (1.9) then it follows that $T = 0$ and hence, $\mathfrak{P}$ is a proper cone.

Example 1.8. Let $\mathcal{H}$ be a complex Hilbert space with inner product $\langle \cdot, \cdot \rangle$. Then $\mathfrak{A} = B(\mathcal{H})$ equipped with the composition as multiplication and normed with the operator norm is a complex Banach algebra. The operators that are self adjoint with respect to the inner product constitute the set of Hermitian elements of the algebra $B(\mathcal{H})$. The elements of the set

$$\mathfrak{P} = \{T \in B(\mathcal{H}) \colon \langle Tx, x \rangle \geqslant 0 \text{ for all } x \in \mathcal{H}\}$$

of nonnegative operators is a cone of nonnegative definite elements of $\mathfrak{A}$, see [BD71, Theorem 8, Section 9].

The following result is rather important for this thesis. It is well known, at least for the bounded operator case as in Example 1.8. The first statement of this Proposition was proved in [Sin71]. The fact that $\sigma(T) \subset \mathbb{V}(T) \subset \mathbb{R}^+$ is a compact set implies the second one.

Proposition 1.9. *Let $T \in \mathfrak{A}$ be an element of the Banach algebra $\mathfrak{A}$. Then the following statements hold.*

(i) *If T is Hermitian, then* $\operatorname{spr}(T) = \operatorname{nr}(T) = \|T\|$;

(ii) *If $T \in \mathfrak{P}$, then* $\operatorname{spr}(T) \in \sigma(T)$.

We are now able to give a more simple representation of ϕ_A under the assumption that the Laurent coefficients of A are nonnegative definite.

Proposition 1.10. *Suppose that A is an $\mathfrak{A}$-valued function which is analytic in its domain $\mathbb{G}$ and suppose that the coefficients $A_j \in \mathfrak{A}$ of its Laurent representation are in the cone $\mathfrak{P} \subset \mathfrak{A}$ of nonnegative definite elements of $\mathfrak{A}$. Then for all $\rho \in [0, \infty) \cap \mathbb{G}$ we have that*

$$\phi_A(\rho) = \operatorname{spr} A(\rho).$$

Proof. Clearly, $\operatorname{spr} A(\rho) \leqslant \phi_A(\rho)$. Hence, only $\sup\limits_{|\lambda|=\rho} \operatorname{spr} A(\lambda) \leqslant \operatorname{spr} A(\rho)$ must be verified.

We have that $A(|\lambda|) \in \mathfrak{P}$ for all $\lambda \in \mathbb{A}_{\eta_1,\eta_2}$ and

$$\begin{aligned}
\operatorname{spr}(A(\lambda)) &\leqslant \operatorname{nr}(A(\lambda)) = \sup\{|f(A(\lambda))| \colon f \in \mathfrak{S}_I(\mathfrak{A}')\} \\
&\leqslant \sup\Big\{\sum_{j\in\mathbb{Z}} |\lambda^j||f(A_j)| \colon f \in \mathfrak{S}_I(\mathfrak{A}')\Big\} \\
&= \sup\Big\{\sum_{j\in\mathbb{Z}} |\lambda^j| f(A_j) \colon f \in \mathfrak{S}_I(\mathfrak{A}')\Big\} \\
&= \sup\Big\{f(A(|\lambda|) \colon f \in \mathfrak{S}_I(\mathfrak{A}')\Big\} \\
&= \operatorname{nr}(A(|\lambda|)) = \operatorname{spr}(A(|\lambda|)),
\end{aligned}$$

where the last equality holds due to Proposition 1.9. □

ϕ_A has the same simple representation if the Laurent coefficients A_j of A are elements of any so-called normal algebra cone of $\mathfrak{A}$, which we will define now.

We will call a subset $\mathfrak{C} \subset \mathfrak{A}$ an **algebra cone** if it satisfies the following conditions

$$\begin{aligned}
&(i) \quad \mathfrak{C} + \mathfrak{C} \subset \mathfrak{C}, && (ii) \quad \lambda\mathfrak{C} \subset \mathfrak{C} \quad \text{for all } \lambda \geqslant 0, \\
&(iii) \quad I \in \mathfrak{C}, && (iv) \quad \mathfrak{C} \cdot \mathfrak{C} \subset \mathfrak{C}.
\end{aligned}$$

It is well-known that for the cone $\mathfrak{P}$ of nonnegative definite elements, Property (iv) does not hold, see e.g. [Con85].

If $\mathfrak{A}$ has an algebra cone $\mathfrak{C}$, we call the pair $(\mathfrak{A}, \mathfrak{C})$ an **ordered Banach algebra**, which is due to the fact that $\mathfrak{C}$ induces an ordering "$\leqslant$" on $\mathfrak{A}$ by

$$T \leqslant S \quad \text{if and only if} \quad S - T \in \mathfrak{C}.$$

An algebra cone $\mathfrak{C}$ of $\mathfrak{A}$ is called **normal** if there exists a constant $\alpha > 0$ such that $\|T\| \leqslant \alpha\|S\|$ for all $S, T \in \mathfrak{A}$ with $0 \leqslant T \leqslant S$.

Suppose that $\mathfrak{C}$ is a normal algebra cone in $\mathfrak{A}$ and take $T \in -\mathfrak{C} \cap \mathfrak{C}$, i.e. $0 \leqslant T \leqslant 0$. So $\|T\| \leqslant \alpha\|0\| = 0$, which implies that a normal algebra cone is automatically proper.

Example 1.11. Let $\mathfrak{A} = \mathbb{C}^{n,n}$ be the set of all complex $n \times n$ matrices with the normal matrix multiplication and equipped with any matrix norm $\|\cdot\|$ such that $\|I_n\| = 1$. Recall that a **matrix norm** is a function from $\mathbb{C}^{n,n}$ to $\mathbb{R}_+$ that satisfies (i) $\|A\| \geqslant 0$ and $\|A\| = 0 \Leftrightarrow A = 0$, (ii) $\|\mu A\| = |\mu|\|A\|$ $\forall \mu \in \mathbb{C}$, (iii) $\|A + B\| \leqslant \|A\| + \|B\|$ and (iv) $\|AB\| \leqslant \|A\|\|B\|$ for all $A, B \in \mathbb{C}^{n,n}$.

Then $(\mathbb{C}^{n,n}, \|\cdot\|)$ is a complex ordered Banach algebra and the subset

$$\mathfrak{C} = \{A = (a_{ij})_{i,j=1}^n \colon a_{ij} \geqslant 0 \text{ for all } i, j \in \langle n \rangle\} \subset \mathbb{C}^{n,n},$$

where $\langle n \rangle := \{1, \ldots, n\}$, is closed and it is a normal algebra cone, see also e.g. [KLS89].

For $A, B \in \mathbb{C}^{n,n}$, $A = (a_{ij})_{i,j=1}^n$, $B = (b_{ij})_{i,j=1}^n$, we write $A \leqslant B$, if $B - A \in \mathfrak{C}$. Note that $A \leqslant B$ is equivalent to $a_{ij} \leqslant b_{ij}$ for all $i, j \in \langle n \rangle$. Furthermore, we write $A < B$ if $a_{ij} < b_{ij}$ for all $i, j \in \langle n \rangle$.

The next Proposition can be found e.g. in [RR96].

Proposition 1.12. *If $\mathfrak{A}$ is a complex ordered Banach algebra with closed normal algebra cone $\mathfrak{C}$, then for all $T \in \mathfrak{C}$ we have* $\operatorname{spr}(T) \in \sigma(T)$.

In order to prove that ϕ_A has the same representation as in Proposition 1.10, if the coefficients of A are in a normal algebra cone, we state one more lemma. The proof is a modification of the proof of [Bon55, Lemma 3] and [FN05b, Lemma 3.1].

Lemma 1.13. *Let $\mathfrak{C} \subset \mathfrak{A}$ be a normal algebra cone in $\mathfrak{A}$. Then there exists a positive constant γ such that for an analytic function $F : \mathbb{G} \to \mathfrak{A}$ with $F(\lambda) = \sum_{j\in\mathbb{Z}} \lambda^j A_j$ as in (1-4) with coefficients $A_j \in \mathfrak{C}$ $(j \in \mathbb{Z})$ we have*

$$\|F(\lambda)\| \leqslant \gamma \|F(|\lambda|)\|$$

for all $\lambda \in \mathbb{G}$.

Proof. Since $\mathfrak{C}$ is a normal cone, there exists a positive constant α such that $B_1, B_2 \in \mathfrak{C}$ and $B_1 \leqslant B_2$ imply that $\|B_1\| \leqslant \alpha\|B_2\|$. We first prove that if $C \in \mathfrak{C}$ and $B \in \mathfrak{A}$ with $-C \leqslant B \leqslant C$ then $\|B\| \leqslant 2\alpha\|C\|$. Indeed, from $C - B \geqslant 0$ and $C + B \geqslant 0$ it follows that

$$\begin{aligned} 2\alpha\|C\| &= \alpha\|C + B + (C - B)\| \\ &\geqslant \max\{\|C + B\|, \|C - B\|\} \\ &\geqslant \frac{1}{2}\left(\|C + B\| + \|C - B\|\right) \\ &= \frac{1}{2}\left(\|C + B\| + \|B - C\|\right) \\ &\geqslant \|B\|. \end{aligned} \tag{1-6}$$

Set $\lambda = re^{i\varphi}$ for some $r > 0$ and some $\varphi \in [0, 2\pi)$ and fix any $\vartheta \in [0, 2\pi)$. Then

$$\begin{aligned} \|F(\lambda)\| = \|e^{i\vartheta}F(\lambda)\| &= \left\| e^{i\vartheta}\sum_{j=0}^{\infty} \lambda^j A_j + e^{i\vartheta}\sum_{j=1}^{\infty} \lambda^{-j} A_{-j} \right\| \\ &= \left\| \sum_{j=0}^{\infty} r^j e^{i(\vartheta+j\varphi)} A_j + \sum_{j=1}^{\infty} r^{-j} e^{i(\vartheta-j\varphi)} A_{-j} \right\| \\ &\leqslant \left\| \sum_{j=0}^{\infty} r^j \cos(\vartheta + j\varphi) A_j + \sum_{j=1}^{\infty} r^{-j} \cos(\vartheta - j\varphi) A_{-j} \right\| \\ &\qquad + \left\| \sum_{j=0}^{\infty} r^j \sin(\vartheta + j\varphi) A_j + \sum_{j=1}^{\infty} r^{-j} \sin(\vartheta - j\varphi) A_{-j} \right\| \\ &\leqslant 2 \sup_{\omega_1,\omega_2\in[0,2\pi]} \left\| \sum_{j=0}^{\infty} r^j \cos(\omega_1 + j\varphi) A_j + \sum_{j=1}^{\infty} r^{-j} \cos(\omega_2 - j\varphi) A_{-j} \right\|. \end{aligned}$$

Since for $j \geqslant 0$

$$-r^j A_j \leqslant r^j \cos(\omega_1 + j\varphi) A_j \leqslant r^j A_j,$$

and for $j \geqslant 1$,

$$-r^{-j} A_{-j} \leqslant r^{-j} \cos(\omega_2 - j\varphi) A_{-j} \leqslant r^{-j} A_{-j},$$

by adding these inequalities, it follows that

$$-\sum_{j\in\mathbb{Z}} r^j A_j \leqslant \sum_{j=0}^{\infty} r^j \cos(\omega_1 + j\varphi)A_j + \sum_{j=1}^{\infty} r^{-j}\cos(\omega_2 - j\varphi)A_j \leqslant \sum_{j\in\mathbb{Z}} r^j A_j \in \mathfrak{C}.$$

Therefore, using (1-6), we get

$$\|F(\lambda)\| \leqslant 4\alpha \Big\| \sum_{j\in\mathbb{Z}} r^j A_j \Big\| = 4\alpha \|F(|\lambda|)\|.$$

□

Note that due to Lemma 1.13 we also have $\|F(\lambda)^k\| \leqslant \gamma\|F(|\lambda|)^k\|$ with the same constant γ, since $F(\cdot)^k$ is still a function analytic in $\mathbb{G}$ with coefficients in $\mathfrak{A}$.

Proposition 1.14. *Suppose that A is an $\mathfrak{A}$-valued function which is analytic in its domain $\mathbb{G}$ and suppose that the coefficients $A_j \in \mathfrak{A}$ of its Laurent representation are in a closed normal algebra cone $\mathfrak{C} \subset \mathfrak{A}$ of $\mathfrak{A}$. Then for all $\rho \in [0,\infty) \cap \mathbb{G}$ we have that*

$$\phi_A(\rho) = \operatorname{spr} A(\rho).$$

Proof. Clearly, $\operatorname{spr} A(\rho) \leqslant \phi_A(\rho)$. Hence, only $\max_{|\lambda|=\rho} \operatorname{spr} A(\lambda) \leqslant \operatorname{spr} A(\rho)$ must be verified. By Lemma 1.13 we have that

$$\phi_A(\rho) = \sup_{|\lambda|=\rho} \operatorname{spr} A(\lambda) = \sup_{|\lambda|=\rho} \lim_{k\to\infty} \|A(\lambda)^k\|^{1/k} \leqslant \sup_{|\lambda|=\rho} \lim_{k\to\infty} \left(\gamma\|A(|\lambda|)^k\|\right)^{1/k} = \operatorname{spr} A(\rho),$$

since $\gamma^{1/k} \to 1$ when $k \to \infty$. □

1.2 Matrix valued functions

In this section we study some spectral properties of an m–monic function F for the special case where $\mathfrak{A} = \mathbb{C}^{n,n}$, i. e., the coefficients A_j of the function (1-5)

$$F(\lambda) = \lambda^m I - \sum_{j\in\mathbb{Z}} \lambda^j A_j, \ \lambda \in \mathbb{G}, \tag{1-7}$$

are complex $n \times n$-matrices for all $j \in \mathbb{Z}$. The spectrum $\sigma(F)$ of F consists only of eigenvalues. If $\lambda \in \mathbb{C}$ is an eigenvalue of F, then obviously λ is a root of the analytic function $\lambda \mapsto \det F(\lambda)$, $\lambda \in \mathbb{G}$. Hence, clearly, F can have all $\lambda \in \mathbb{C}$ as eigenvalues. We call the order of λ as a root of $\det F$ the **multiplicity** of the eigenvalue λ.

The first proposition of this section gives necessary conditions in terms of the function ϕ_A for disks around zero to contain a certain number of eigenvalues of F. These conditions will be useful later.

Proposition 1.15. *Let F be as in* (1-7).

(*i*) *Let $\rho \in (0,\infty) \cap \mathbb{G}$ be such that $\phi_A(\rho) < \rho^m$. Then F has exactly mn eigenvalues (counting multiplicities) in $\mathbb{D}_\rho$.*

(*ii*) *Let $\rho \in (0,\infty) \cap \mathbb{G}$ and $\delta > 0$ such that $\phi_A(\tau) < \tau^m$ when $\tau \in (\rho, \rho+\delta) \subset \mathbb{G}$. Then F has exactly mn eigenvalues (counting multiplicities) in the closed disc $\overline{\mathbb{D}}_\rho$.*

(*iii*) *Let* $\rho \in (0, \infty) \cap \mathbb{G}$ *and* $\rho > \delta > 0$ *such that* $\phi_A(\tau) < \tau^m$ *when* $\tau \in (\rho - \delta, \rho)$. *Then* F *has exactly* mn *eigenvalues (counting multiplicities) in* $\mathbb{D}_\rho$.

Proof. (*i*) Let $\epsilon \in [0, 1]$ and consider the function F_ϵ with $F_\epsilon(\lambda) = \lambda^m I_n - \epsilon A(\lambda)$. For any $\epsilon \in [0, 1]$ the matrix $F_\epsilon(\lambda)$ is invertible for all $\lambda \in \mathbb{T}_\rho$; indeed

$$\operatorname{spr}(\epsilon A(\lambda)) \leqslant \operatorname{spr} A(\lambda) \leqslant \sup_{|\lambda|=\rho} \operatorname{spr} A(\lambda) = \phi_A(\rho) < \rho^m = |\lambda|^m.$$

Therefore, since the eigenvalues of F_ϵ continuously depend on ϵ, the number of eigenvalues of F_ϵ in $\mathbb{D}_\rho$ is the same for every $\epsilon \in [0, 1]$. $F_0(\lambda) = \lambda^m I_n$ has mn eigenvalues in the disc $\mathbb{D}_\rho$ and so has $F_1(\lambda) = F(\lambda)$.

(*ii*) By (*i*), F has mn eigenvalues in $\mathbb{D}_\tau$ for $\rho < \tau < \rho+\delta$, i. e., in $\bigcap_{\tau \in (\rho, \rho+\delta)} \mathbb{D}_\tau = \overline{\mathbb{D}}_\rho$.

(*iii*) Analogously, by (*i*), F has mn eigenvalues in $\bigcup_{\tau \in (\rho-\delta, \rho)} \mathbb{D}_\tau = \mathbb{D}_\rho$.

□

We now briefly look at the restrictions on the principal part of A, which follow if the function ϕ_A, which is closely associated with the spectral radii of the matrices $A(\tau)$ $(\tau \geqslant 0)$, is bounded.

Proposition 1.16. *Let* $\mathbb{G} = \mathbb{C}$ *and* A *as in* (1-4) *with* $A_j \in \mathbb{C}^{n,n}$ *for all* $j \in \mathbb{N}$. *Suppose that* ϕ_A *is bounded. Then for* $j \in \mathbb{N}$, $j \neq 0$ *the following statements hold.*

(*i*) *If* A_j *is entrywise nonnegative, then* A_j *is nilpotent.*

(*ii*) *If* A_j *is nonnegative definite, then* $A_j = 0$.

Proof. ϕ_A is bounded so let $M \in \mathbb{R}^+$ be such that $\operatorname{spr} A(\rho) = \phi_A(\rho) \leqslant M$ for all $\rho \in [0, \infty)$.

(*i*) If $A_j \geqslant 0$, then we have $\rho^j A_j \leqslant A(\rho)$, i. e. $A_j \leqslant \rho^{-j} A(\rho)$ for all $\rho > 0$. Hence,

$$\operatorname{spr} A_j \leqslant \rho^{-j} \operatorname{spr} A(\rho) \leqslant \rho^{-j} M \overset{\rho\to\infty}{\longrightarrow} 0.$$

Therefore, $\operatorname{spr} A_j = 0$, i. e., A_j is nilpotent.

(*ii*) For $v, w \in \mathbb{C}^n$ denote by $\langle v, w\rangle = w^* v$ the standard inner product in $\mathbb{C}^n$.

If A_j is positive semidefinite, then, due to Proposition 1.9(*i*), for all $j \in \mathbb{N}\backslash\{0\}$ and for all $v \in \mathbb{C}^n$ with $\|v\| = 1$ we have that

$$\rho^j \langle A_j v, v\rangle \leqslant \sum_{j\in\mathbb{Z}} \rho^j \langle A_j v, v\rangle = \langle A(\rho)v, v\rangle \leqslant \operatorname{nr} A(\rho) = \operatorname{spr} A(\rho) \leqslant M,$$

i. e.,

$$\langle A_j v, v\rangle \leqslant \rho^{-j} M$$

for all $\rho > 0$. Analogously to (*i*) it follows that $\langle A_j v, v\rangle = 0$ for all $v \in \mathbb{C}^n$ with $\|v\| = 1$, hence, due to Proposition 1.9, $\|A_j\| = 0$, i. e., $A_j = 0$.

□

Remark 1.17. *Note that Proposition 1.16(ii) also holds if the coefficients A_j are bounded operators in a Hilbert space. The proof works in exactly the same way.*

The next example shows that the reverse direction of Proposition 1.16(i) does not hold in general.

Example 1.18. Let $\mathfrak{A} = \mathbb{C}^{2,2}$ and

$$A(\lambda) = \begin{bmatrix} 0 & \lambda \\ \lambda^2 & 0 \end{bmatrix},$$

then A_1 and A_2 are entrywise nonnegative and nilpotent. But, since $\sigma(A(\lambda)) = \{\pm\lambda^{3/2}\}$, it follows that $\phi_A(\rho) = \rho^{3/2}$, which is not bounded.

Proposition 1.19. *Let A be as in (1-4) with entrywise nonnegative coefficients $A_j \in \mathbb{C}^{n,n}$. Suppose that there are $\rho_1, \rho_2 \in \mathbb{R} \cap \mathbb{G}$, $0 < \rho_1 \leqslant \rho_2$ such that $\phi_A(\rho_j) = \rho_j^m$ for $j = 1, 2$. Then we have that*

(i) *$\phi_A(\rho) = \rho^m$ for all $\rho \in [\rho_1, \rho_2]$ if and only if $\sigma(F) = \mathbb{G}$;*

(ii) *$\phi_A(\rho) < \rho^m$ for all $\rho \in (\rho_1, \rho_2)$ if and only if $\sigma(F) \cap \mathbb{A}_{\rho_1,\rho_2} = \varnothing$.*

Proof. (i) If $\phi_A(\rho) = \rho^m$ for all $\rho \in [\rho_1, \rho_2]$, then by Proposition 1.14, $\operatorname{spr} A(\rho) = \rho^m$. $A(\rho)$ is a nonnegative matrix. By the well known Perron-Frobenius theorem, see e.g. [BP94], [HJ85], [Min88], the spectral radius of $A(\rho)$ is an eigenvalue of $A(\rho)$. Hence, $\det(\rho^m I_n - A(\rho)) = 0$ for all $\rho \in [\rho_1, \rho_2]$. A is an analytic matrix function in $\mathbb{G}$, hence, $\det(\lambda^m I_n - A(\lambda)) = 0$ for all $\lambda \in \mathbb{G}$.

If $\sigma(F) = \mathbb{G}$ holds, then for every $\rho \in [\rho_1, \rho_2]$, ρ^m is an eigenvalue of $A(\rho)$, hence, $\phi_A(\rho) = \operatorname{spr} A(\rho) \geqslant \rho^m$ for all $\rho \in [\rho_1, \rho_2]$ and therefore case (ii) from Proposition 1.4 cannot occur, thus, $\phi_A(\rho) = \rho^m$ for all $\rho \in [\rho_1, \rho_2]$.

(ii) The 'only if' part is precisely Proposition 1.7. For the 'if' part suppose that $\sigma(F) \cap \mathbb{A}_{\rho_1,\rho_2} = \varnothing$. The assertion then follows from (i) and Proposition 1.4.

□

The following lemma can be found in [HJ91, p.491, formula (6.5.9)]. Here we adopt it to our setting.

Lemma 1.20. *Let $F : \mathbb{R} \to \mathbb{C}^{n,n}$ be a differentiable matrix function. Then*

$$(\det F)'(\tau) = \sum_{j=1}^{n} (\det F_{(j)})(\tau), \tag{1-8}$$

where $F_{(j)}(\tau)$ is the matrix that coincides with $F(\tau)$ except that the j-th column is differentiated with respect to τ.

Proposition 1.21. *Let $F(\lambda) = \lambda^m I_n - A(\lambda)$ be as in (1-5) and suppose that for all $\tau \geqslant 0$ we have $\phi_A(\tau) \in \sigma(A(\tau))$. Then the following statements hold.*

(i) *Let ϕ_A be differentiable in $\rho \in \mathbb{G} \cap \mathbb{R}$, $\rho > 0$. Then $\phi_A(\rho) = \rho^m$ implies that*

$$(\det F)'(\rho) = \left(m\rho^{m-1} - \phi_A'(\rho)\right) \operatorname{tr}\left[\operatorname{adj}(F(\rho)\right].$$

(*ii*) *Let* ϕ_A *be twice differentiable in* $\rho \in \mathbb{G} \cap \mathbb{R},\ \rho > 0$. *Then* $\phi_A(\rho) = \rho^m$ *and* $\phi'_A(\rho) = m\rho^{m-1}$ *imply that*

$$(\det F)''(\rho) = \big(m(m-1)\rho^{m-2} - \phi''_A(\rho)\big)\operatorname{tr}\left[\operatorname{adj}(F(\rho)\right].$$

Proof. Define the function $G : \mathbb{R}^+ \to \mathbb{C}^{n,n},\ \tau \mapsto \phi_A(\tau)I_n - A(\tau)$.

In order to simplify the notation, in this proof we introduce the following abbreviations. We write F and G column wise as

$$F(\rho) = \Big[F_1(\rho), \ldots, F_n(\rho)\Big] \text{ and } G(\rho) = \Big[G_1(\rho), \ldots, G_n(\rho)\Big],$$

with $F_j(\rho), G_j(\rho) \in \mathbb{C}^n,\ j \in \langle n \rangle$ and denote by $e_j \in \mathbb{C}^n$ the j-th unit vector.

For $v, w \in \mathbb{C}^n$, $\tau > 0$ and $j \neq k$ set

$$D_j^{(\tau)}(v) = \sum_{j=1}^{n} \det\Big[F_1(\tau), \ldots, F_{j-1}(\tau), v, F_{j+1}(\tau), \ldots, F_n(\tau)\Big]$$

and

$$D_{jk}^{(\tau)}(v,w) = \begin{cases} \det\Big[F_1(\tau), \ldots, F_{j-1}(\tau), v, F_{j+1}(\tau), \ldots, \\ \qquad\qquad F_{k-1}(\tau), w, F_{k+1}(\tau), \ldots, F_n(\tau)\Big],\ j < k \\ \\ \det\Big[F_1(\tau), \ldots, F_{k-1}(\tau), w, F_{k+1}(\tau), \ldots, \\ \qquad\qquad F_{j-1}(\tau), v, F_{j+1}(\tau), \ldots, F_n(\tau)\Big],\ j > k \end{cases}.$$

(*i*) Since $\phi_A(\tau) \in \sigma(A(\tau))$ for all $\tau \geqslant 0$, the function $\det G : \mathbb{R}^+ \to \mathbb{R}$ is constantly zero, hence, differentiable with $(\det G)' \equiv 0$.

By assumption we have $F(\rho) = G(\rho)$. Furthermore, if we denote by $(A'(\rho))_j$ the derivative of the jth column of $A(\rho)$ with respect to ρ, $F'_j(\rho) = m\rho^{m-1}e_j - (A'(\rho))_j$ and $G'_j(\rho) = \phi'_A(\rho)e_j - A'_j(\rho)$ implies

$$F'_j(\rho) = G'_j(\rho) + \big(m\rho^{m-1} - \phi'_A(\rho)\big)e_j \quad (j \in \langle n \rangle). \tag{1-9}$$

Then due to Lemma 1.20

$$\begin{aligned} 0 = (\det G)'(\rho) &= \sum_{j=1}^{n} \det\Big[G_1(\rho), \ldots, G_{j-1}(\rho), G'_j(\rho), G_{j+1}(\rho), \ldots, G_n(\rho)\Big] \\ &= \sum_{j=1}^{n} D_j^{(\rho)}(G'_j(\rho)) \end{aligned}$$

and therefore

$$\begin{aligned} (\det F)'(\rho) =& \sum_{j=1}^{n} D_j^{(\rho)}(F'_j(\rho)) = \sum_{j=1}^{n} D_j^{(\rho)}(G'_j(\rho)) + D_j^{(\rho)}\big(\big(m\rho^{m-1} - \phi'_A(\rho)\big)e_j\big) \\ =& \big(m\rho^{m-1} - \phi'_A(\rho)\big)\sum_{j=1}^{n} D_j^{(\rho)}(e_j) = \big(m\rho^{m-1} - \phi'_A(\rho)\big)\sum_{j=1}^{n}\big[\operatorname{adj} F(\rho)\big]_{jj} \\ =& \big(m\rho^{m-1} - \phi'_A(\rho)\big)\operatorname{tr}\left[\operatorname{adj}(F(\rho)\right]. \end{aligned}$$

(*ii*) Note that by assumption (1-9) implies $F_j'(\rho) = G_j'(\rho)$ for $j \in \langle n \rangle$. Furthermore,

$$F_j''(\rho) = G_j''(\rho) + (m(m-1)\rho^{m-2} - \phi_A''(\rho))e_j.$$

Differentiating $(\det F)'$ and considering the derivative at ρ we obtain

$$\begin{aligned}
(\det F)''(\rho) =& \sum_{j=1}^{n} \frac{d}{d\tau} D_j^{(\tau)} \left(F_j'(\tau)\right)\bigg|_{\tau=\rho} \\
=& \sum_{\substack{j,k=1 \\ j\neq k}}^{n} D_{jk}^{(\rho)} \left(F_j'(\rho), F_k'(\rho)\right) + \sum_{j=1}^{n} D_j^{(\rho)} \left(F_j''(\rho)\right) \\
=& \sum_{\substack{j,k=1 \\ j\neq k}}^{n} D_{jk}^{(\rho)} \left(G_j'(\rho), G_k'(\rho)\right) + \sum_{j=1}^{n} D_j^{(\rho)} \left(G_j''(\rho)\right) \\
&+ \left(m(m-1)\rho^{m-2} - \phi_A''(\rho)\right) \sum_{j=1}^{n} D_j^{(\rho)}(e_j) \\
=& (\det G)''(\rho) + \left(m(m-1)\rho^{m-2} - \phi_A''(\rho)\right) \sum_{j=1}^{n} D_j^{(\rho)}(e_j),
\end{aligned}$$

where the last equality follows from $F(\rho) = G(\rho)$. Since already $(\det G)' \equiv 0$, as was shown in (*i*), and thus $(\det G)'' \equiv 0$, we have

$$\begin{aligned}
(\det F)''(\rho) =& \left(m(m-1)\rho^{m-2} - \phi_A''(\rho)\right) \sum_{j=1}^{n} D_j^{(\rho)}(e_j) \\
=& \left(m(m-1)\rho^{m-2} - \phi_A''(\rho)\right) \sum_{j=1}^{n} [\operatorname{adj} F(\rho)]_{jj} \\
=& \left(m(m-1)\rho^{m-2} - \phi_A''(\rho)\right) \operatorname{tr} [\operatorname{adj} F(\rho)] .
\end{aligned}$$

□

Note that in the case that the matrix coefficients A_j of the function F in (1-5) are nonnegative definite matrices and also in the case that the A_j are entrywise nonnegative, the conditions of Proposition 1.21 are met. This follows from Proposition 1.9 and Proposition 1.12, respectively.

1.3 Matrix polynomials

In Chapter 4 and 5 we will study special matrix functions, namely polynomials, i. e., functions P with values in $\mathfrak{A} = \mathbb{C}^{n,n}$ that can be written as

$$P(\lambda) = \sum_{j=0}^{l} \lambda^j A_j, \tag{1-10}$$

where $A_j \in \mathbb{C}^{n,n}$ for all $j \in \langle l \rangle_0$. Notice that P is analytic on the whole complex plane, so $\mathbb{G}$ can be chosen as $\mathbb{C}$.

In the following we will recall some notation and definitions concerning the spectrum of matrix polynomials in general.

A matrix polynomial P is called **regular** if the scalar polynomial $\det P(\lambda)$ is not identically zero. It is called **singular** if it is not regular. P is called **unimodular** if $\det P(\lambda)$ is constant nonzero for all $\lambda \in \mathbb{C}$.

Two matrix polynomials P_1 and P_2 are called **equivalent** if there are unimodular matrix polynomials E and F such that

$$P_2(\lambda) = E(\lambda)P_1(\lambda)F(\lambda)$$

for all $\lambda \in \mathbb{C}$. We write $P_1 \sim P_2$.

Let P be a matrix polynomial with $P(\lambda) = \sum_{j=0}^{l} \lambda^j A_j$. The scalar polynomial $\det P(\lambda)$ is called the **characteristic polynomial** of P. The set of its roots coincides with the (finite) spectrum $\sigma(P)$ of P, which consists only of eigenvalues. The **algebraic multiplicity** $a(P, \lambda_0)$ of an eigenvalue λ_0 of P is the multiplicity of λ_0 as a root of $\det P(\lambda)$. The **geometric multiplicity** $g(P, \lambda_0)$ is the dimension of the kernel $\mathsf{N}(P(\lambda_0))$ of $P(\lambda_0)$. If P is regular, the degree $\deg \det P$ is not greater than ln, hence the number of eigenvalues of P counting algebraic multiplicities does not exceed ln.

Next we introduce the notion of Jordan chains and Jordan pairs for a finite eigenvalue as it is done in [GLR82]. For $k \in \mathbb{N}$ denote by $P^{(k)}(\lambda)$ the k-th derivative $P^{(k)}(\lambda) = \sum_{j=k}^{l} \frac{j!}{(j-k)!} \lambda^{j-k} A_j$ of P with respect to λ.

Let $\lambda_0 \in \mathbb{C}$ be an eigenvalue of a matrix polynomial P and let $x_0, x_1, \ldots, x_\kappa$, be a sequence of n-dimensional vectors with $x_0 \neq 0$, for which the identities

$$\sum_{j=0}^{\nu} \frac{1}{j!} P^{(j)}(\lambda_0) x_{\nu-j} = 0, \quad \nu = 0, ..., \kappa$$

hold. Then $x_0, x_1, \ldots, x_\kappa$ is called a **Jordan chain** of length $\kappa + 1$ for P corresponding to the eigenvalue λ_0. Setting $\nu = 0$ shows that x_0 is an eigenvector of P corresponding to λ_0.

Let P be a matrix polynomial as in (1-10) and suppose it is regular. Let $\lambda_0 \in \mathbb{C}$ be an eigenvalue of P and let

$$x_0^{(i)}, \ldots, x_{\kappa_i - 1}^{(i)}, \quad i = 1, \ldots, \gamma_0 = g(P, \lambda_0)$$

be a set of Jordan chains for P corresponding to λ_0. Then this set is called a **canonical set of Jordan chains** for P if the eigenvectors $x_0^{(1)}, \ldots, x_0^{(\gamma_0)}$ are linearly independent and $\sum_{i=1}^{\gamma_0} \kappa_i = a(P, \lambda_0)$. We call the lengths of the Jordan chains of a canonical set of Jordan chains the **partial multiplicities** of P at λ_0.

Note that arbitrary sets of Jordan chains of matrix polynomials are in general neither canonical nor can they be prolonged to become canonical. Consider for instance

$$P(\lambda) = \lambda I - \left(\lambda^2 \begin{bmatrix} -1 & 0 \\ 0 & -1 \end{bmatrix} + \lambda \begin{bmatrix} 1 & 1 \\ 0 & 1 \end{bmatrix} \right) = \begin{bmatrix} \lambda^2 & -\lambda \\ 0 & \lambda^2 \end{bmatrix}.$$

We have $\det P(\lambda) = \lambda^4$, so 0 is the only eigenvalue and, since $P(0) = 0$, each nonzero vector from $\mathbb{C}^2$ is an eigenvector of P corresponding to 0. Now $P'(0) = \left[\begin{smallmatrix} 0 & -1 \\ 0 & 0 \end{smallmatrix}\right]$ and $P''(0) = \left[\begin{smallmatrix} 2 & 0 \\ 0 & 2 \end{smallmatrix}\right]$, so $x_0, x_1 = \left[\begin{smallmatrix} 1 \\ 0 \end{smallmatrix}\right], \left[\begin{smallmatrix} 1 \\ 0 \end{smallmatrix}\right]$ and $y_0 = \left[\begin{smallmatrix} 0 \\ 1 \end{smallmatrix}\right]$ form a set of Jordan chains corresponding to the eigenvalue 0. Suppose that x_0, x_1, x_2 with $x_2 \in \mathbb{C}^2$ is a Jordan chain corresponding to 0. This implies that $P''(0)\left[\begin{smallmatrix}1\\0\end{smallmatrix}\right] + P'(0)\left[\begin{smallmatrix}1\\0\end{smallmatrix}\right] = 0$, or $\left[\begin{smallmatrix}2\\0\end{smallmatrix}\right] = 0$. If we suppose that y_0, y_1 with $y_1 \in \mathbb{C}^2$ is a Jordan chain, then $P'(0)\left[\begin{smallmatrix}0\\1\end{smallmatrix}\right] = 0$ or $\left[\begin{smallmatrix}-1\\0\end{smallmatrix}\right] = 0$. Hence, the Jordan chains cannot be prolonged and we have $\kappa_1 + \kappa_2 = 2 + 1 < a(P, 0)$.

The next important theorem and its proof can be found in [GLR82, Chapter S1] for the general case of $m \times n$ matrix polynomials. We formulate it for the case of $n \times n$ matrix polynomials.

Theorem 1.22. *Let P be any matrix polynomial with coefficients in $\mathbb{C}^{n,n}$. Then there exist unimodular $n \times n$ matrix polynomials E and F such that*

$$P(\lambda) = E(\lambda)D(\lambda)F(\lambda),$$

where

$$D(\lambda) = \begin{bmatrix} d_1(\lambda) & & & & & \\ & \ddots & & & & \\ & & d_r(\lambda) & & & \\ & & & 0 & & \\ & & & & \ddots & \\ & & & & & 0 \end{bmatrix}$$

and d_i, $i = 1, \ldots, r$, are monic scalar polynomials such that d_i is divisible by d_{i-1} for $i = 2, \ldots, r$.

Note that Theorem 1.22 also holds for singular matrix polynomials. The diagonal of $D(\lambda)$ contains zeros if and only if P is singular.

The matrix polynomial D in Theorem 1.22 is called the **Smith form** of P and the polynomials d_i are called the **invariant polynomials** of the matrix polynomial P. The invariant polynomials d_i are of the form

$$d_i(\lambda) = (\lambda - \lambda_1)^{\alpha_{i1}} \cdots (\lambda - \lambda_{\kappa_i})^{\alpha_{i\kappa_i}},$$

where $\lambda_1, \ldots, \lambda_{\kappa_i}$ are mutually different and $\alpha_{i1}, \ldots, \alpha_{i\kappa_i}$ are positive integers. The factors $(\lambda - \lambda_j)^{\alpha_{ij}}$, $j \in \langle\, \kappa_i \,\rangle$ $i \in \langle\, r \,\rangle$ are called **elementary divisors** of P.

One can easily read off the number and lengths of Jordan chains in a canonical set of Jordan chains for each eigenvalue of P from the Smith form. The number of elementary divisors corresponding to an eigenvalue λ_0 appearing in the Smith form coincides with the number of Jordan chains corresponding to λ_0 and therewith also gives the dimension of the eigenspace $\mathsf{N}(P(\lambda_0))$. Furthermore, the degree of each elementary divisor gives the length of the corresponding Jordan chain, see e.g. [GLR82]. For illustration we give a simple example.

Consider the regular matrix polynomial

$$P(\lambda) = \begin{bmatrix} \lambda^2-\lambda & 1 & 2\lambda^2-2\lambda-2 & 0 & 2\lambda^3-2\lambda^2+\lambda \\ \lambda-1 & 0 & -5\lambda^2+5\lambda & \lambda^6+\lambda^5-2\lambda^4 & \lambda^6-\lambda^5-3\lambda^4+3\lambda^2 \\ 0 & 0 & \lambda^2-\lambda & 0 & \lambda^3-\lambda^2 \\ \lambda^3-\lambda^2 & \lambda & 2\lambda^3-2\lambda^2-2\lambda & \lambda^4+\lambda^3-2\lambda^2 & 2\lambda^4-2\lambda^3+\lambda^2 \\ -\lambda+1 & 0 & 5\lambda^2-5\lambda & -\lambda^6-\lambda^5+2\lambda^4 & 5\lambda^3-5\lambda^2 \end{bmatrix}$$

of degree $l = 6$. Its Smith form turns out to be

$$D(\lambda) = \begin{bmatrix} 1 & & & & \\ & \lambda-1 & & & \\ & & \lambda(\lambda-1) & & \\ & & & \lambda^2(\lambda-1)(\lambda+2) & \\ & & & & \lambda^2(\lambda-1)^3(\lambda+2) \end{bmatrix}.$$

Therefore, the finite spectrum of P consists of $\{\lambda_1 = -2, \lambda_2 = 0, \lambda_3 = 1\}$ with $a(P,-2) = g(P,-2) = 2$, $a(P,0) = 5$, $g(P,0) = 3$ and $a(P,1) = 6$, $g(P,1) = 4$. There are two Jordan chains corresponding to λ_1, each of length one. Corresponding to λ_2, there are three Jordan chains, two of them with length two and one with length one. Finally, there are four Jordan chains corresponding to the third eigenvalue, λ_3. One has length three and the remaining three have lengths one, respectively.

We introduce the following notation. If for $k \geqslant 1$, $(X_j)_{j=0}^{k-1}$ is any sequence of $\mu \times \nu$-matrices, we denote by $\mathbf{col}(X_j)_{j=0}^{k-1}$ the $k \times 1$ block matrix $\begin{bmatrix} X_0 \\ \vdots \\ X_{k-1} \end{bmatrix} \in \mathbb{C}^{k\mu,\nu}$.

Let (X, J) be a pair of matrices, where X is an $n \times \mu$ matrix and J is a $\mu \times \mu$ Jordan matrix with only eigenvalue $\lambda_0 \in \mathbb{C}$. We call (X, J) a **Jordan pair** of P corresponding to the eigenvalue λ_0 if the following conditions hold.

1. λ_0 is a zero of $\det P(\lambda)$ with multiplicity μ,
2. $\operatorname{rank} \operatorname{col}(XJ^j)_{j=0}^{l-1} = \mu$, with $\operatorname{col}(XJ^j)_{j=0}^{l-1} \in \mathbb{C}^{ln,\mu}$, $\mu \leqslant ln$,
3. $A_l X J^l + A_{l-1} X J^{l-1} + \cdots + A_0 X = 0$.

Notice that in [GLR82] the notion of a Jordan pair is defined in a different way. Theorem 7.1 in the same book states that the three conditions which we used to define a Jordan pair are necessary and sufficient for (X, J) to be a Jordan pair for λ_0. Suppose the Jordan matrix J is of the form

$$J = \begin{bmatrix} J_1 & & & \\ & J_2 & & \\ & & \ddots & \\ & & & J_d \end{bmatrix},$$

with $d \leqslant \mu$ and J_j is a $\kappa_j \times \kappa_j$ Jordan block corresponding to λ_0 for $j = 1, \ldots, d$. Then, we have $\sum_{j=1}^{d} \kappa_j = \mu$.

Again, let P be a matrix polynomial as in (1-10). The matrix polynomial $\operatorname{rev} P$ with

$$\operatorname{rev} P(\lambda) = \lambda^l A_0 + \cdots + \lambda A_{l-1} + A_l = \lambda^l P(1/\lambda)$$

is called the **reverse matrix polynomial** of P. We say that P has an **eigenvalue at infinity** if 0 is an eigenvalue of $\operatorname{rev} P$. The notions of algebraic and geometric multiplicities, Jordan chains, canonical sets of Jordan chains, invariant polynomials, elementary divisors and Jordan pairs for infinite eigenvalues can be introduced in a completely analogous and obvious way.

Consider a regular matrix polynomial P of degree l. Then the scalar polynomial $\det(P(\lambda))$ has at most degree nl. Let us suppose that the degree of $\det(P(\lambda))$ is d, i. e., it has precisely d roots, counting multiplicities. Now consider

$$\det(\operatorname{rev} P(\lambda)) = \det(\lambda^l P(1/\lambda)) = \lambda^{ln} \det(P(1/\lambda)).$$

Therefore, the determinant of the reverse polynomial $\operatorname{rev} P$ has zero as a root exactly with multiplicity $nl - d$. Hence, we can state the following remark.

Remark 1.23. *If P is a regular $n \times n$-matrix polynomial of degree l, it always has exactly nl eigenvalues, counting finite and infinite eigenvalues each with their multiplicities.*

A major part of this thesis is dedicated to m–monic matrix polynomials

$$P(\lambda) = \lambda^m I_n - \sum_{j=0}^{l} \lambda^j A_j = \lambda^m I_n - A(\lambda),$$

where $n, m, l \in \mathbb{N}$, $m < l$, $A(\lambda) = \sum_{j=0}^{l} \lambda^j A_j$ and $A_j \in \mathbb{C}^{n,n}$ for $j \in \langle l \rangle_0$.

Clearly, it is possible to write every matrix polynomial in m–monic form. This form becomes interesting when the coefficients A_j are supposed to have special properties. In Chapter 4 and 5 we study m–monic matrix polynomials such that the coefficients A_j are entrywise nonnegative or even irreducible nonnegative.

In some occasions we have to deal with monic matrix polynomials such that all but the highest order coefficient are entrywise nonpositive. The following simple consequence of [FN91] will turn out useful, so we present it here.

Recall that for entrywise nonnegative matrices $A, B \geqslant 0$ with $A \leqslant B$ it follows that $\operatorname{spr}(A) \leqslant \operatorname{spr}(B)$, see e. g. [BP94], [HJ85], [Min88].

Proposition 1.24. *Let P be a monic matrix polynomial with*

$$P(\lambda) = \lambda^l I_n - A(\lambda) = \lambda^l I_n - \sum_{j=0}^{l-1} \lambda^j A_j, \tag{1-11}$$

where the coefficients A_j are entrywise nonnegative matrices for $j \in \langle\, l-1 \,\rangle_0$. Let $0 < r = \operatorname{spr} P < \infty$ be the spectral radius of P. Then

$$r^l = \operatorname{spr}(A(r)).$$

Furthermore, for $\rho > 0$, we have that $\operatorname{spr} A(\rho) < \rho^l$ if and only if $r < \rho$.

Proof. Due to the main theorem in [FN91] we have

$$r = \operatorname{spr}(\operatorname{rev} A(1/r)) = \operatorname{spr}\left(A_{l-1} + \frac{1}{r} A_{l-2} + \cdots + \frac{1}{r^{l-2}} A_1 + \frac{1}{r^{l-1}} A_0 \right).$$

Multiplying this equation with r^{l-1} proves the first assertion.

Let $\alpha_1, \ldots, \alpha_{l-1}$ be such that $A_j \leqslant \alpha_j \mathbf{1}_{n,n}$ for $j \in \langle\, l-1 \,\rangle_0$, where $\mathbf{1}_{n,n}$ denotes the matrix with all entries equal to 1. Note that $\operatorname{spr}(\mathbf{1}_{n,n}) = n$. Then we have for $\tau > 0$

$$\operatorname{spr} A(\tau) = \operatorname{spr}\left(\sum_{j=0}^{l-1} \tau^j A_j \right) \leqslant \sum_{j=0}^{l-1} \tau^j \alpha_j \operatorname{spr}(\mathbf{1}_{n,n}) = \sum_{j=0}^{l-1} n \alpha_j \tau^j.$$

Therefore, for the function $\phi_A : 0 < \tau \mapsto \operatorname{spr} A(\tau)$ we have that $\phi_A(\tau) = \mathcal{O}(\tau^{l-1})$, i. e. $\limsup_{\tau\to\infty} \left| \frac{\phi_A(\tau)}{\tau^{l-1}} \right| < \infty$. Furthermore, there exists at most one $\tau > 0$ such that $\operatorname{spr}(A(\tau)) = \tau^l$, see [FN91]. Hence, the last assertion follows. □

For the next result, recall that a nonsingular **M-matrix** is an invertible **Z-matrix** (i. e. a matrix the off-diagonal entries of which are nonpositive), such that its inverse is entrywise nonnegative. The next well known lemma can be found in e.g. [BP94].

Lemma 1.25. *A Z–matrix is a nonsingular M–matrix if and only if it can be written as $\tau I - B$, where $\tau > 0$ and B is an entrywise nonnegative matrix such that* $\operatorname{spr}(B) < \tau$.

Proposition 1.26. *Let P be a monic matrix polynomial with*

$$P(\lambda) = \lambda^l I_n - A(\lambda) = \lambda^l I_n - \sum_{j=0}^{l-1} \lambda^j A_j,$$

where the coefficients A_j are nonnegative matrices for $j \in \langle l-1 \rangle$ and let $\operatorname{spr}(P) = 0$. *Then for all $\tau > 0$ the matrix $P(\tau)$ is a nonsingular M-matrix.*

Proof. Suppose there exists a $\tau > 0$ such that $\operatorname{spr} A(\tau) = \phi_A(\tau) \geqslant \tau^l$. Then, since $\phi_A(\tau) = \mathcal{O}(\tau^{l-1})$, there exists a $\rho \geqslant \tau$ such that $\phi_A(\rho) = \rho^l$, which implies that ρ is an eigenvalue of P. This is a contradiction to $\operatorname{spr}(P) = 0$. □

Monic polynomials as in (1-11) have been thoroughly studied for instance in [PT04], [Rau92].

2 m–monic operator functions which are analytic on an annulus with self adjoint coefficients

In [Wim08] it was shown that if monic matrix polynomials of the form $\lambda^m I - \sum_{j=0}^{m-1} \lambda^j A_j$, where $A_0, \dots, A_{m-1}$ are self adjoint matrices, have eigenvalues on the unit circle, then they are rotation invariant with respect to the angles of certain roots of unity, if the condition $\sum_{j=0}^{m-1} |A_j| \leqslant I$ is satisfied. In [SW10], this result was extended to monic operator polynomials with bounded nonnegative definite operator coefficients.

In this chapter we extend the results of J. Swoboda and H. K. Wimmer to functions F with values in the Banach algebra $B(\mathcal{H})$ of bounded operators on a Hilbert space $\mathcal{H}$ which are analytic on an annulus and can be written as $F(\lambda) = \lambda^m I - \sum_{j \in \mathbb{Z}} \lambda^j A_j$, where the coefficients A_j are self adjoint. We will show that some eigenvalues of F are distributed like certain roots of unity on circles corresponding to their modulus.

2.1 Preliminaries

The setting in this chapter is the following.

(a) $\mathcal{H}$ is a Hilbert space, with inner product $\langle .\,, . \rangle$;

(b) For $-\infty < \rho_0 < \rho_2 \leqslant \infty$, $\rho_2 > 0$ set

$$\mathbb{A}_{\rho_0, \rho_2} = \{\lambda \in \mathbb{C} : \rho_0 < |\lambda| < \rho_2\}$$

and $\rho_1 = \max\{0, \rho_0\}$;

(c) Let $m \in \mathbb{Z}$, and let F be a function with values in $B(\mathcal{H})$ which is analytic in $\mathbb{A}_{\rho_0, \rho_2}$ and can be written as

$$F(\lambda) = \lambda^m I - A(\lambda) = \lambda^m I - \sum_{j \in \mathbb{Z}} \lambda^j A_j \quad \text{with } A_j \in B(\mathcal{H}),\ j \in \mathbb{Z}; \tag{2-12}$$

(d) $A_j = A_j^*$ for $j \in \mathbb{Z}$.

Note that if $\rho_0 < 0$, then $\mathbb{A}_{\rho_0, \rho_2} = \mathbb{D}_{\rho_2}$. In this case we have for F that $A_{-j} = 0$ for all $j \in \mathbb{N} \setminus \{0\}$, i. e. the Laurent representation of F has only a regular part.

For $T, S \in B(\mathcal{H})$, we will write $T \geqslant 0$ if T is nonnegative definite in the sense of Example 1.8 and more generally, $T \geqslant S$ or $S \leqslant T$ if $T - S \geqslant 0$. If T is nonnegative definite, denote by $T^{1/2}$ the unique nonnegative definite operator in $B(\mathcal{H})$ such that $(T^{1/2})^2 = T$. $T^{1/2}$ is called the nonnegative **square root** of T, see e.g. [Kat76], [Wer00].

In this section we always suppose that the following assumption is satisfied.

Assumption 2.1. *There exists a $\rho \in (\rho_1, \rho_2)$ such that*

$$|A|(\rho) := \sum_{j \in \mathbb{Z}} \rho^j |A_j| \leqslant \rho^m I, \tag{2-13}$$

where $|A_j| = (A_j^2)^{1/2}$ for $j \in \mathbb{Z}$.

Define the function $F_{|A|} : \mathbb{A}_{\rho_0,\rho_2} \to B(\mathcal{H})$ via

$$F_{|A|}(\lambda) = \lambda^m I - |A|(\lambda) = \lambda^m I - \sum_{j\in\mathbb{Z}} \lambda^j |A_j|. \tag{2-14}$$

The next simple observation will be used frequently in what follows.

Lemma 2.2. *For $T \in \mathcal{B}(\mathcal{H})$, $T^* = T$ and $v \in \mathcal{H}$, $v \neq 0$ the inequality*

$$|\langle Tv, v\rangle| \leqslant \langle |T|v, v\rangle$$

holds and therefore $|T| - T$ is a nonnegative definite operator.

Proof. This follows immediately from the spectral theorem (see any standard textbook on functional analysis, e.g. [Heu06], [Wer00], [Kre89], [Con85], [Kat76]). Denote by E_λ the spectral function of T. Then

$$|\langle Tv, v\rangle| = \left|\int_{\mathbb{R}} \lambda \, d\langle E_\lambda v, v\rangle\right| \leqslant \int_{\mathbb{R}} |\lambda| \, d\langle E_\lambda v, v\rangle = \langle |T|v, v\rangle .$$

□

As in Chapter 1, consider the geometrically convex function $\phi_{|A|} : (\rho_1, \rho_2) \to [0, \infty)$,

$$\phi_{|A|}(\tau) = \sup_{|\lambda|=\tau} \operatorname{spr} |A|(\lambda).$$

The next corollary gives a representation of $\phi_{|A|}$ and follows directly from Proposition 1.10.

Corollary 2.3. *For all $\rho \in (\rho_1, \rho_2)$ we have $\phi_{|A|}(\rho) = \operatorname{spr} |A|(\rho)$.*

We will study the consequences of Assumption 2.1 and the property of $\phi_{|A|}$ being geometrically convex on (ρ_1, ρ_2) for the spectrum F in Section 2.2. Let us first take a brief look on the connection between Assumption 2.1 and the function $\phi_{|A|}$.

Let Assumption 2.1 hold. Then due to Proposition 1.9 we have

$$\phi_{|A|}(\rho) = \operatorname{spr} |A|(\rho) = \| |A|(\rho)\| \leqslant \rho^m, \tag{2-15}$$

where the last inequality is due to (2-13).

If conversely $\phi_{|A|}(\rho) \leqslant \rho^m$ holds, then for all $v \in \mathcal{H}$

$$\langle |A|(\rho)v, v\rangle \leqslant \| |A|(\rho)\| \|v\|^2 = \phi_{|A|}(\rho)\|v\|^2 \leqslant \rho^m \|v\|^2 = \langle \rho^m v, v\rangle , \tag{2-16}$$

which means that $|A|(\rho) \leqslant \rho^m I$.

If even $\phi_{|A|}(\rho) < \rho^m$ is satisfied, then similar to (2-16) for $v \neq 0$

$$\langle |A|(\rho)v, v\rangle < \langle \rho^m v, v\rangle .$$

Obviously if $|A|(\rho) = \rho^m I$ then $\phi_{|A|}(\rho) = \rho^m$. Conversely, if $\phi_{|A|}(\rho) = \rho^m$ then by (2-16) there must be a $v \in \mathcal{H}$, $v \neq 0$ such that $\langle |A|(\rho)v, v\rangle = \rho^m \|v\|^2$. Hence, $\langle F_{|A|}(\rho)v, v\rangle = 0$, where due to (2-16), $F_{|A|}(\rho) = \rho^m I - |A|(\rho)$ is a nonnegative definite operator. Therefore, $\| \left(F_{|A|}(\rho)\right)^{1/2} v\| = 0$, hence, $F_{|A|}(\rho)v = 0$ or $|A|(\rho)v = \rho^m v$.

Summarizing these considerations we obtain the following statement.

Lemma 2.4. *For $\rho \in (\rho_1, \rho_2)$,*

$$\phi_{|A|}(\rho) \leqslant \rho^m \quad \text{if and only if} \quad |A|(\rho) \leqslant \rho^m I$$

and if $\phi_{|A|}(\rho) < \rho^m$, then

$$\langle\, |A|(\rho)v, v\rangle < \langle \rho^m v, v\rangle \quad \text{for all } v \in \mathcal{H},\ v \neq 0.$$

Furthermore, if $\phi_{|A|}(\rho) = \rho^m$, then there is a $v \in \mathcal{H}, v \neq 0$ such that $|A|(\rho)v = \rho^m v$, i. e., ρ is an eigenvalue of $F_{|A|}$ with eigenvector v.

Due to Lemma 2.4, the general Assumption 2.1 is equivalent to $\phi_{|A|}(\rho) \leqslant \rho^m$ for some $\rho \in (\rho_1, \rho_2)$. If $\phi_{|A|}(\rho) < \rho^m$, there is a maximal interval $I = (\rho_\alpha, \rho_\omega) \subseteq (\rho_1, \rho_2)$ with $\phi_{|A|}(\tau) < \tau^m$ for all $\tau \in I$. Because of Proposition 1.4 we have $\phi_{|A|}(\tau) > \tau^m$ for all $\tau \in (\rho_1, \rho_2) \setminus [\rho_\alpha, \rho_\omega]$. So there is precisely one open interval in (ρ_1, ρ_2) (its boundary points can coincide with ρ_1 or ρ_2) on which $\phi_{|A|}(\rho) = \operatorname{spr}|A|(\rho) < \rho^m$ holds. By Proposition Proposition 1.7, $F_{|A|}$ has no spectrum in $\mathbb{A}_{\rho_\alpha, \rho_\omega}$.

From $\rho_\alpha, \rho_\omega \in (\rho_1, \rho_2)$, it follows that $\phi_{|A|}(\rho_\alpha) = \rho_\alpha^m$ and $\phi_{|A|}(\rho_\omega) = \rho_\omega^m$. Lemma 2.4 then implies that ρ_α and ρ_ω are eigenvalues of $F_{|A|}$.

2.2 Rotation invariance of eigenvalues

In this section we study the distribution of eigenvalues of F on circles of radius ρ satisfying $\phi_{|A|}(\rho) = \rho^m$. We will see that they have a certain rotation invariance property.

For $v \in \mathcal{H}$, $v \neq 0$ and $\rho \in (\rho_1, \rho_2)$ let us introduce two sets of numbers which we will make frequently use of.

(a) $\mathbb{I}(v) = \{j \in \mathbb{Z} \setminus \{m\} \colon \langle A_j v, v\rangle \neq 0\}$,

(b) $\mathbb{S}(\rho, v) = \{\lambda \in \mathbb{C} : \langle F(\lambda)v, v\rangle = 0,\ |\lambda| = \rho\}$.

Lemma 2.5. *Let F be as in (2-12). Let $\rho \in (\rho_1, \rho_2)$ satisfy Assumption 2.1. Let $\lambda \in \mathbb{S}(\rho, v)$ for some $v \in \mathcal{H}$, $v \neq 0$. Then we have*

(i) $\left(\frac{\rho}{\lambda}\right)^{m-j} \langle A_j v, v\rangle = \langle |A_j| v, v\rangle$ *for all* $j \in \mathbb{Z}$;

(ii) $\left(\frac{\rho}{\lambda}\right)^{m-j} = \pm 1$ *for all* $j \in \mathbb{I}(v)$.

Proof. Let w.l.o.g. $\|v\| = 1$. We then have $\lambda^m = \sum\limits_{j\in\mathbb{Z}} \lambda^j \langle A_j v, v\rangle$. By multiplying with $\frac{\rho^m}{\lambda^m}$ we obtain

$$\rho^m = \sum_{j\in\mathbb{Z}} \rho^j \left(\frac{\rho}{\lambda}\right)^{m-j} \langle A_j v, v\rangle = \sum_{j\in\mathbb{I}(v)\cup\{m\}} \rho^j \left(\frac{\rho}{\lambda}\right)^{m-j} \langle A_j v, v\rangle \tag{2-17}$$

$$= \left| \sum_{j\in\mathbb{I}(v)} \rho^j \left(\frac{\rho}{\lambda}\right)^{m-j} \langle A_j v, v\rangle \right| \leqslant \sum_{j\in\mathbb{I}(v)} \rho^j \left|\langle A_j v, v\rangle\right| \tag{2-18}$$

$$\leqslant \sum_{j\in\mathbb{I}(v)} \rho^j \langle |A_j| v, v\rangle \leqslant \sum_{j\in\mathbb{Z}} \rho^j \langle |A_j| v, v\rangle \leqslant \rho^m, \tag{2-19}$$

so in (2-18) and (2-19) we have equality.

Recall that if for $\alpha_j \in \mathbb{C}$ ($j \in \mathbb{Z}$) the identity $\sum_{j\in\mathbb{Z}} \alpha_j = \sum_{j\in\mathbb{Z}} |\alpha_j|$ holds, then we have $\alpha_j = |\alpha_j|$ for all $j \in \mathbb{Z}$.

Due to (2-17)-(2-19) we have that

$$\sum_{j\in\mathbb{Z}} \rho^j \left(\frac{\rho}{\lambda}\right)^{m-j} \langle A_j v, v\rangle = \sum_{j\in\mathbb{Z}} \left|\rho^j \left(\frac{\rho}{\lambda}\right)^{m-j} \langle A_j v, v\rangle\right| = \sum_{j\in\mathbb{Z}} \rho^j \langle |A_j| v, v\rangle ,$$

which implies (i).

Since $\lambda \in \mathbb{S}(\rho, v)$, we have that $\left|\frac{\rho}{\lambda}\right| = 1$. Thus from (i) and $\langle A_j v, v\rangle \in \mathbb{R}$, (ii) follows immediately. □

Corollary 2.6. *Consider F be as in* (2-12). *Let $\mathbb{S}(\rho, v) \neq \varnothing$ for some $v \in \mathcal{H}$, $v \neq 0$ and let $\rho \in (\rho_1, \rho_2)$ satisfy Assumption 2.1. Then for all $j \in \mathbb{Z}$ the following statements are equivalent.*

(i) $\langle A_j v, v\rangle = 0$; (ii) $A_j v = 0$;

(iii) $\langle |A_j| v, v\rangle = 0$; (iv) $|A_j| v = 0$.

Proof. The equivalence of (i) and (iii) follows immediately from Lemma 2.5(i).

To prove the remaining equivalences, observe that

$$\langle |A_j| v, v\rangle = 0 \quad \text{iff} \quad \||A_j|^{1/2} v\| = 0 \quad \text{which implies} \quad |A_j| v = 0.$$

Now $0 = |A_j| v = (A_j^2)^{1/2} v$ implies $A_j^2 v = 0$ and thus by the self adjointness of A_j, $A_j v = 0$. Clearly, that implies $\langle A_j v, v\rangle = 0$, which completes the proof. □

Note that, therefore, under the assumptions of Lemma 2.5, the set $\mathbb{I}(v)$ takes the form

$$\mathbb{I}(v) = \{j \in \mathbb{Z} \setminus \{m\} \colon A_j v = 0\}. \tag{2-20}$$

The next proposition shows that under a certain condition on the coefficients A_j, $\lambda \in \mathbb{C}$ is an eigenvalue of F if and only if $|\lambda|$ is an eigenvalue of $F_{|A|}$.

Proposition 2.7. *Let F be as in* (2-12) *and $F_{|A|}$ as in* (2-14). *Let $\rho \in (\rho_1, \rho_2)$ satisfy Assumption 2.1. Then for $v \in \mathcal{H} \setminus \{0\}$ and $\lambda \in \mathbb{C}$ the following statements are equivalent.*

(i) *$\lambda \in \mathbb{S}(\rho, v)$;*

(ii) *$F(\lambda)v = 0$ and $|\lambda| = \rho$;*

(iii) *$F_{|A|}(\rho)v = 0$ and we have $A_j v = \left(\frac{\lambda}{\rho}\right)^{m-j} |A_j| v$ for all $j \in \mathbb{Z}$.*

Proof. <u>$(ii) \Rightarrow (i)$</u>: This follows immediately from the definition of $\mathbb{S}(\rho, v)$.

<u>$(i) \Rightarrow (iii)$</u>: By Lemma 2.5, $\left(\frac{\rho}{\lambda}\right)^{m-j} \langle A_j v, v\rangle = \langle |A_j| v, v\rangle$ for all $j \in \mathbb{Z}$, hence

$$\left\langle \left(|A_j| - \left(\frac{\rho}{\lambda}\right)^{m-j} A_j \right) v, v \right\rangle = 0 \quad \text{for all } j \in \mathbb{Z}. \tag{2-21}$$

Since $\left|\left(\frac{\rho}{\lambda}\right)^{m-j} A_j\right| = |A_j|$, Lemma 2.2 implies that the operators $\Delta_j = |A_j| - \left(\frac{\rho}{\lambda}\right)^{m-j} A_j$ ($j \in \mathbb{Z}$) are nonnegative operators. So (2-21) implies that $\|\Delta_j^{1/2} v\| = 0$, hence, $\Delta_j v = 0$ and therefore

$$|A_j|v = \left(\frac{\rho}{\lambda}\right)^{m-j} A_j v \quad \text{for all } j \in \mathbb{Z}, \tag{2-22}$$

which is precisely the second identity in (iii).

(2-19) implies that $\left\langle \left(\rho^m I - \sum_{j\in\mathbb{Z}} \rho^j |A_j|\right) v, v\right\rangle = 0$, where due to the assumption that $\phi_{|A|}(\rho) \leqslant \rho^m$, the operator $\rho^m I - \sum_{j\in\mathbb{Z}} \rho^j |A_j|$ is nonnegative definite, hence, by the square root argument again,

$$\rho^m v = \sum_{j\in\mathbb{Z}} \rho^j |A_j| v, \tag{2-23}$$

and (iii) is proved.

$(iii) \Rightarrow (ii)$: This immediately follows from

$$\rho^m v = \sum_{j\in\mathbb{Z}} \rho^j |A_j| v = \sum_{j\in\mathbb{Z}} \rho^j \left(\frac{\rho}{\lambda}\right)^{m-j} A_j v,$$

i. e., $\lambda^m v = \sum_{j\in\mathbb{Z}} \lambda^j A_j v$.

□

Corollary 2.8. *Let $\rho \in (\rho_1, \rho_2)$ satisfy Assumption 2.1, and let $v \in \mathcal{H}$ $v \neq 0$. Then the following assertions hold.*

(i)
$$\mathbb{S}(\rho, v) = \{\lambda \in \mathbb{C} : F(\lambda)v = 0,\ |\lambda| = \rho\},$$
i. e., $\mathbb{S}(\rho, v)$ consists of all eigenvalues of F with modulus ρ, for which v is an eigenvector.

(ii) *If $\mathbb{S}(\rho, v) \neq \varnothing$ for some $\rho > 0$, then we have that $\phi_{|A|}(\rho) = \rho^m$.*

(iii) *$\lambda \in \mathbb{S}(\rho, v)$ if and only if $\bar{\lambda} \in \mathbb{S}(\rho, v)$.*

Proof. (i) This is a direct consequence of Proposition 2.7.

(ii) From (i) and Proposition 1.7, it follows that if $\phi_{|A|}(\rho) < \rho^m$, then ρ is no spectral point of $F_{|A|}$ and due to Proposition 2.7, $\mathbb{S}(\rho, v)$ is empty for all $v \in \mathcal{H}$.

(iii) Due to $F(\lambda)^* = F(\bar{\lambda})$, we have $0 = \langle F(\lambda)v, v\rangle = \langle F(\bar{\lambda})v, v\rangle$, hence, $\bar{\lambda} \in \mathbb{S}(\rho, v)$.

□

The aim for the remainder of this section is to study the geometry of the set $\mathbb{S}(\rho, v)$.

For $k \in \mathbb{Z}$ denote by $\mathbb{E}_k \subset \mathbb{T} \subset \mathbb{C}$ the set of the k-th roots of unity, i. e., the set

$$\mathbb{E}_k = \{z \in \mathbb{C} : z^k = 1\} = \{e^{i\frac{j}{k}2\pi} : j \in \langle\, k\,\rangle\}.$$

For $z \in \mathbb{E}_k$ denote by $\operatorname{ord} z$ the **order** of z, i. e., the smallest divisor s of k such that $z^s = 1$. Note that $\mathbb{E}_k$ constitutes a multiplicative group with unit 1 and inverse element $(e^{i\frac{j}{k}2\pi})^{-1} = e^{i\frac{k-j}{k}2\pi}$.

Lemma 2.9. *Let $\rho \in (\rho_1, \rho_2)$ satisfy Assumption 2.1. For $v \in \mathcal{H} \setminus \{0\}$ and $j \in \mathbb{I}(v)$ the following statements hold.*

(i) $\mathbb{S}(\rho, v) \subset \rho\,\mathbb{E}_{2(m-j)}$.

(ii) $\mathbb{S}(\rho, v) \subset \rho\,\mathbb{E}_{m-j}$ *if* $\rho \in \mathbb{S}(\rho, v)$.

Proof. Take $\lambda \in \mathbb{S}(\rho, v)$ and $j \in \mathbb{I}(v)$. By Proposition 2.7, we have $A_j v = \left(\frac{\lambda}{\rho}\right)^{m-j} |A_j| v$, and hence,

$$\left(\frac{\rho}{\lambda}\right)^{m-j} \langle A_j v, v\rangle = \langle |A_j| v, v\rangle > 0.$$

By Lemma 2.5, $\left(\frac{\rho}{\lambda}\right)^{m-j} = \pm 1$, i. e. $\left(\frac{\lambda}{\rho}\right)^{2(m-j)} = 1$, in other words $\frac{\lambda}{\rho} \in \mathbb{E}_{2(m-j)}$. This proves (i). If furthermore $\rho \in \mathbb{S}(\rho, v)$, then due to Proposition 2.7 we have

$$\rho^{m-j} A_j v = \lambda^{m-j} |A_j| v \quad \text{and} \quad \rho^{m-j} A_j v = \rho^{m-j} |A_j| v,$$

thus, $(\lambda^{m-j} - \rho^{m-j})|A_j| v = 0$. From $A_j v \neq 0$ and Corollary 2.6 follows that $|A_j| v \neq 0$. Since $j \neq m$, we get $\left(\frac{\lambda}{\rho}\right)^{m-j} = 1$ and (ii) follows. □

Lemma 2.10. *Let F be as in (2-12) and let $\rho \in (\rho_1, \rho_2)$ satisfy Assumption 2.1. For $v \in \mathcal{H}$, $v \neq 0$, let $\lambda \in \mathbb{S}(\rho, v)$ with $\operatorname{ord} \frac{\lambda}{\rho} = k$. Then the following statements hold.*

(i) If k is odd, then

(a) $k \,|(m-j)$ *for all* $j \in \mathbb{I}(v)$ *(i. e., $j = m + \nu k$ for some $\nu \in \mathbb{Z}$.).*

(b) $A_j v = |A_j| v$ *for all* $j \in \mathbb{Z}$.

(ii) If k is even, i.e., $k = 2s$, then

(a) $s\,|(m-j)$ *for all* $j \in \mathbb{I}(v)$ *(i. e., $j = m + \nu s$ for some $\nu \in \mathbb{Z}$.)*

(b) $A_j v = (-1)^\nu |A_j| v$ *for all $j \in \mathbb{I}(v)$, where $\nu \in \mathbb{Z}$ s.t. $j = m + \nu s$.*

Proof. (i) If $k = \operatorname{ord} \frac{\lambda}{\rho}$ is odd, then $\left(\frac{\lambda}{\rho}\right)^\nu \neq -1$ for all $\nu \in \mathbb{Z}$. Indeed, suppose that $\left(\frac{\lambda}{\rho}\right)^\nu = -1$. Then $k \mid (2\nu)$ and, since k is odd, $k \mid \nu$. But this implies $\left(\frac{\lambda}{\rho}\right)^\nu = 1$, which contradicts the assumption. So, since due to Lemma 2.5, for all $j \in \mathbb{I}(v)$ we have $\left(\frac{\lambda}{\rho}\right)^{m-j} = \pm 1$, it follows that $\left(\frac{\lambda}{\rho}\right)^{m-j} = 1$, and therefore $k\,|m-j$ for all $j \in \mathbb{I}(v)$.

Equation (b) clearly holds for $j \notin \mathbb{I}(v)$. Using Proposition 2.7 and (a), we immediately obtain $A_j v = |A_j| v$ for $j \in \mathbb{I}(v)$.

(ii) (a) is clear, this is Lemma 2.9.

Let $j \in \mathbb{I}(v)$ and $j = m - \nu s$ for some $\nu \in \mathbb{Z}$. By Proposition 2.7

$$A_j v = \left(\frac{\lambda}{\rho}\right)^{m-j} |A_j| v = \left(\frac{\lambda}{\rho}\right)^{\nu s} |A_j| v.$$

Since $\left(\frac{\lambda}{\rho}\right)^k = \left(\frac{\lambda}{\rho}\right)^{2s} = 1$ and k is the smallest integer s.t. $\left(\frac{\lambda}{\rho}\right)^k = 1$, we have $\left(\frac{\lambda}{\rho}\right)^s = -1$. Hence, $A_j = (-1)^\nu |A_j| v$.

□

Corollary 2.11. *Let F be as in* (2-12) *and let $\rho \in (\rho_1, \rho_2)$ satisfy Assumption 2.1 and fix some nonzero $v \in \mathcal{H}$. Suppose that $\mathbb{S}(\rho, v) \neq \varnothing$ and $\mathbb{I}(v) \neq \varnothing$. Then the following statements are equivalent.*

(*i*) *There exists $\lambda \in \mathbb{S}(\rho, v)$ such that $k = \operatorname{ord} \frac{\lambda}{\rho}$ is odd*

(*ii*) $A_j v = |A_j| v$ *for all* $j \in \mathbb{Z}$

(*iii*) $\rho \in \mathbb{S}(\rho, v)$.

Proof. (*i*) $\Rightarrow$ (*ii*): This is Lemma 2.10(*i*).
(*ii*) $\Rightarrow$ (*iii*): Take $\lambda \in \mathbb{S}(\rho, v)$. Then by Proposition 2.7

$$\rho^m v = \sum_{j \in \mathbb{Z}} \rho^j |A_j| v = \sum_{j \in \mathbb{Z}} \rho^j A_j v,$$

i. e., $\rho \in \mathbb{S}(\rho, v)$.
(*iii*) $\Rightarrow$ (*i*) is clear, since $\operatorname{ord} \frac{\rho}{\rho} = 1$. □

Lemma 2.12. *Let F be as in* (2-12) *and let $\rho \in (\rho_1, \rho_2)$ satisfy Assumption 2.1. Suppose that $\rho \in \mathbb{S}(\rho, v)$ for some $v \in \mathcal{H}$, $v \neq 0$ and let $d \in \mathbb{Z}$. Then the following statements are equivalent.*

(*i*) $\mathbb{I}(v) \subset \{m + kd \colon k \in \mathbb{Z}\}$.

(*ii*) $F(\lambda)v = \lambda^m(\lambda^d - \rho^d)\theta(\lambda^d)$ *with $\theta : \mathbb{C} \to \mathcal{H}$ such that the function $\lambda \mapsto \theta(\lambda^d)$ is analytic in $\mathbb{A}_{\rho_1, \rho_2}$.*

(*iii*) $\rho\, \mathbb{E}_d \subset \mathbb{S}(\rho, v)$.

(*iv*) *There exists a $\lambda \in \mathbb{S}(\rho, v)$ such that $\operatorname{ord} \frac{\lambda}{\rho} = d$.*

Proof.

(*i*) $\Rightarrow$ (*iii*): Take $\lambda \in \rho\, \mathbb{E}_d$, i. e. $\lambda^d = \rho^d$. Then due to (*i*)

$$\begin{aligned} F(\lambda)v &= \left(\lambda^m I - \sum_{j \in \mathbb{Z}} \lambda^j A_j\right) v = \left(\lambda^m I - \sum_{\nu \in \mathbb{Z}} \lambda^{m+\nu d} A_{m+\nu d}\right) v \\ &= \lambda^m \left(I - \sum_{\nu \in \mathbb{Z}} \lambda^{\nu d} A_{m+\nu d}\right) v \\ &= \left(\frac{\lambda}{\rho}\right)^m \left(\rho^m I - \sum_{\nu \in \mathbb{Z}} \rho^m \lambda^{\nu d} A_{m+\nu d}\right) v \qquad (2\text{-}24) \\ &= \left(\frac{\lambda}{\rho}\right)^m \left(\rho^m I - \sum_{\nu \in \mathbb{Z}} \rho^{m+\nu d} A_{m+\nu d}\right) v = \left(\frac{\lambda}{\rho}\right)^m F(\rho) v = 0. \end{aligned}$$

(*iii*) $\Rightarrow$ (*i*): From Lemma 2.9 it follows that $\rho\, \mathbb{E}_d \subset \mathbb{S}(\rho, v) \subset \rho\, \mathbb{E}_{m-j}$ for all $j \in \mathbb{I}(v)$, and hence, $d \mid (m - j)$ for each $j \in \mathbb{I}(v)$, which is equivalent to (*i*).

(*iii*) ⇔ (*ii*): Suppose that (*ii*) holds. Then for each $\lambda \in \mathbb{C}$ such that $\lambda^d = \rho^d$ we have that $F(\lambda)v = 0$, i. e., $\lambda \in \mathbb{S}(\rho, v)$.

Now, suppose that (*iii*) holds. Since we already proved that (*iii*) is equivalent to (*i*), we can write $F(\lambda)v = \lambda^m \tilde{F}(\lambda^d)$. Due to (*iii*), each $\lambda \in \mathbb{C}$ such that $\lambda^d = \rho^d$ is a zero of the analytic function $\lambda \mapsto F(\lambda)v$ mapping from $\mathbb{A}_{\rho_1,\rho_2}$ to $\mathbb{C}^n$. Hence, the assertion (*ii*) follows.

(*iii*) ⇒ (*iv*): Obviously, any $\lambda \in \rho\,\mathbb{E}_d$ will do.

(*iv*) ⇒ (*i*): Due to Lemma 2.9 (*i*) we have that $\lambda \in \rho\,\mathbb{E}_{m-j}$ for all $j \in \mathbb{I}(v)$. From $d = \operatorname{ord} \frac{\lambda}{\rho}$ it follows that $d\,|(m-j)$ for all $j \in \mathbb{I}(v)$.

□

We will now prove the main results of this section.

For a subset $M \subset \mathbb{Z}$ denote by $\gcd(M)$ the greatest common divisor of M, i. e., the greatest integer $\mu \in \mathbb{N}$ such that μ divides all elements of M and by $\operatorname{lcm}(M)$ the least common multiple of M, i. e. the smallest integer $\nu \in \mathbb{N}$ such that ν is a multiple of each element of M.

Theorem 2.13. *Let F be as in* (2-12) *and let $\rho \in (\rho_1, \rho_2)$ satisfy Assumption 2.1, i.e.*

$$\sum_{j\in\mathbb{Z}} \rho^j |A_j| \leqslant \rho^m I.$$

Suppose that $\rho \in \mathbb{S}(\rho, v)$ for some $v \in \mathcal{H}$, $v \neq 0$ and set

$$\begin{aligned}\hat{d} &= \max\{d \in \mathbb{Z} : d\,|(m-j) \quad \textit{for all } j \in \mathbb{I}(v)\} \\ &= \gcd\{m-j : j \in \mathbb{I}(v)\}.\end{aligned}$$

Then $\mathbb{S}(\rho, v) = \rho\,\mathbb{E}_{\hat{d}}$.

Proof. Let $j_0 \in \mathbb{I}(v)$. From Lemma 2.9, we know that $\rho^{-1}\mathbb{S}(\rho, v) \subset \mathbb{E}_{m-j_0}$.

We show that $\rho^{-1}\mathbb{S}(\rho, v)$ is closed under multiplication.

Let $\frac{\lambda_1}{\rho}, \frac{\lambda_2}{\rho} \in \rho^{-1}\mathbb{S}(\rho, v)$ with $\operatorname{ord} \frac{\lambda_i}{\rho} = k_i$, $i = 1, 2$. From Lemma 2.12 it follows that for $i = 1, 2$ and for all $j \in \mathbb{I}(v)$ we have that $k_i\,|(m-j)$, i. e., $A_j v \neq 0$ only if $m - j \in k_1\mathbb{Z} \cap k_2\mathbb{Z} = p\mathbb{Z}$, where $p = \operatorname{lcm}\{k_1, k_2\}$. Hence, $p\,|(m-j)$ for all $j \in \mathbb{I}(v)$ and furthermore $\left(\frac{\lambda_1}{\rho}\frac{\lambda_2}{\rho}\right)^p = 1$, i. e., $\frac{\lambda_1}{\rho}\frac{\lambda_2}{\rho} \in \mathbb{E}_p$. Applying Lemma 2.12 again, it follows that $\mathbb{E}_p \subset \rho^{-1}\mathbb{S}(\rho, v)$, thus $\frac{\lambda_1}{\rho}\frac{\lambda_2}{\rho} \in \rho^{-1}\mathbb{S}(\rho, v)$.

So $\rho^{-1}\mathbb{S}(\rho, v)$ is a multiplicative subgroup of $\mathbb{E}_{m-j_0}$, which means that $\rho^{-1}\mathbb{S}(\rho, v) = \mathbb{E}_{\tilde{d}}$, where $\tilde{d} = \max\{\operatorname{ord} \frac{\lambda}{\rho} : \lambda \in \mathbb{S}(\rho, v)\}$. Once more by Lemma 2.12 we have

$$\{\operatorname{ord} \frac{\lambda}{\rho} : \lambda \in \mathbb{S}(\rho, v)\} = \{d \in \mathbb{Z} : d\,|(m-j) \text{ for all } j \in \mathbb{I}(v)\},$$

so $\tilde{d} = \hat{d}$.

□

Theorem 2.14. *Let $\rho \in (\rho_1, \rho_2)$ satisfy Assumption 2.1, i.e.,*

$$\sum_{j\in\mathbb{Z}} \rho^j |A_j| \leqslant \rho^m I.$$

Suppose that $\mathbb{S}(\rho, v) \neq \varnothing$ for some $v \in \mathcal{H}$, $v \neq 0$ and $\rho \notin \mathbb{S}(\rho, v)$. Then $\operatorname{ord}\frac{\lambda}{\rho}$ is even for all $\lambda \in \mathbb{S}(\rho, v)$. Let

$$\hat{d} = \operatorname{lcm}\{\frac{1}{2}\operatorname{ord}\frac{\lambda}{\rho} : \lambda \in \mathbb{S}(\rho, v)\}.$$

Then $\hat{d}\,|(m-j)$ for all $j \in \mathbb{I}(v)$,

$$\mathbb{S}(\rho, v) = \{\lambda \in \mathbb{C} : \lambda^{\hat{d}} + \rho^{\hat{d}} = 0\}$$

and

$$F(z)v = z^m (z^{\hat{d}} + \rho^{\hat{d}})\theta(z^{\hat{d}}), \tag{2-25}$$

with $\theta : \mathbb{C} \to \mathcal{H}$ such that the function $\lambda \mapsto \theta(\lambda^d)$ is analytic in $\mathbb{A}_{\rho_1,\rho_2}$.

Proof. The statement about $\operatorname{ord}\frac{\lambda}{\rho}$ follows directly from Lemma 2.11.

Let $\mathbb{S}(\rho, v) = \{\lambda_1, \dots, \lambda_r\}$ with $\operatorname{ord}\frac{\lambda_\mu}{\rho} = k_\mu$ and $k_\mu = 2s_\mu$, $\mu = 1, \dots, r$. Since by Lemma 2.9 $k_\mu\,|(2(m-j))$, clearly $s_\mu|(m-j)$ for all $j \in \mathbb{I}(v)$ and for all $\mu = 1, \dots, r$. Hence, $\hat{d}\,|(m-j)$. This means, that from $A_j v \neq 0$ it follows that $j = m + \nu\hat{d}$ for some $\nu \in \mathbb{Z}$ holds true. Therefore, analogously to (2-24), for all $z \in \mathbb{C}$ we get

$$F(z)v = \left(\frac{z}{\rho}\right)^m \left(\rho^m I - \sum_{\nu\in\mathbb{Z}} \rho^m z^{\nu\hat{d}} A_{m+\nu\hat{d}}\right) v. \tag{2-26}$$

Clearly $\left(\frac{\lambda}{\rho}\right)^{\hat{d}} \in \{-1, 1\}$ for all $\lambda \in \mathbb{S}(\rho, v)$. Suppose that for some $\lambda \in \mathbb{S}(\rho, v)$ we have $\left(\frac{\lambda}{\rho}\right)^{\hat{d}} = 1$, i. e., $\lambda^{\hat{d}} = \rho^{\hat{d}}$. Then,

$$\begin{aligned} 0 = F(\lambda)v &= \left(\frac{\lambda}{\rho}\right)^m \left(\rho^m I - \sum_{\nu\in\mathbb{Z}} \rho^m \lambda^{\nu\hat{d}} A_{m+\nu\hat{d}}\right) v \\ &= \left(\frac{\lambda}{\rho}\right)^m \left(\rho^m I - \sum_{\nu\in\mathbb{Z}} \rho^{m+\nu\hat{d}} A_{m+\nu\hat{d}}\right) v \\ &= \left(\frac{\lambda}{\rho}\right)^m F(\rho)v, \end{aligned}$$

which would imply that $\rho \in \mathbb{S}(\rho, v)$, and contradict the assumption.

So $\left(\frac{\lambda_\mu}{\rho}\right)^{\hat{d}} = -1$ for $\mu \in \langle\, r\,\rangle$ i. e., $\mathbb{S}(\rho, v) \subset \{\lambda \in \mathbb{C} : \lambda^{\hat{d}} + \rho^{\hat{d}} = 0\}$.

Suppose now that $\lambda \in \{\lambda \in \mathbb{C} : \lambda^{\hat{d}} + \rho^{\hat{d}} = 0\}$. This means that $\lambda^{\hat{d}} = -\rho^{\hat{d}} = \lambda_\mu^{\hat{d}}$ for $\mu \in \langle\, r\,\rangle$. Then

$$0 = F(\lambda_\mu)v = \lambda_\mu^m I - \sum_{\nu\in\mathbb{Z}} \lambda_\mu^{m+\nu\hat{d}} A_{m+\nu\hat{d}} v = \lambda_\mu^m \left(I - \sum_{\nu\in\mathbb{Z}} \lambda_\mu^{\nu\hat{d}} A_{m+\nu\hat{d}}\right) v.$$

Multiplying this equation with $\left(\frac{\lambda}{\lambda_\mu}\right)^m$ and using $\lambda^{\hat{d}} = \lambda_\mu^{\hat{d}}$, we obtain

$$0 = \lambda^m I - \sum_{\nu\in\mathbb{Z}} \lambda^{m+\nu\hat{d}} A_{m+\nu\hat{d}} v = F(\lambda)v.$$

Hence, $\lambda \in \mathbb{S}(\rho, v)$ and therefore $\mathbb{S}(\rho, v) = \{\lambda \in \mathbb{C} : \lambda^{\hat{d}} + \rho^{\hat{d}} = 0\}$.

(2-25) follows as in the proof of Lemma 2.12. □

2.3 m–monic operator polynomials with Hermitian coefficients

Consider now the case that F is an m–monic operator polynomial $F(\lambda) = \lambda^m I - A(\lambda)$ such that A is of degree $l < m$, i. e., F is monic. If $\rho \in (\rho_1, \rho_2)$ satisfies Assumption 2.1 with $\mathbb{S}(\rho, v) \neq \varnothing$ for some nonzero $v \in \mathcal{H}$, then the eigenvalues of F on $\mathbb{T}_\rho$ are always simple. For the matrix case this has been proved in [Wim08]. For the operator case this is also true. Indeed, suppose that $\mu \in \mathbb{S}(\rho, v)$ for some nonzero $v \in \mathcal{H}$ and that there exists a $w \in \mathcal{H}$ such that v, w is part of a Jordan chain of F corresponding to μ, i. e. we have $F'(\mu)v + F(\mu)w = 0$. Multiplying both sides of this identity with v, we obtain with Corollary 2.8

$$0 = \langle F'(\mu)v, v\rangle + \langle F(\mu)w, v\rangle = \langle F'(\mu)v, v\rangle + \langle w, F(\bar{\mu})v\rangle = \langle F'(\mu)v, v\rangle .$$

This can be written as

$$\sum_{j=1}^{l} j\mu^{j-1} \langle A_j v, v\rangle = m\mu^{m-1}\|v\|^2,$$

or due to Proposition 2.7(iii) and $m > j$ for all $j \in \langle l\rangle_0$,

$$\begin{aligned} \|v\|^2 &= \sum_{j=1}^{l} \frac{j}{m}\mu^{j-m} \langle A_j v, v\rangle = \sum_{j=1}^{l} \frac{j}{m}\rho^{j-m} \langle |A_j| v, v\rangle \\ &= \sum_{j=0}^{l} \frac{j}{m}\rho^{j-m} \langle |A_j| v, v\rangle < \sum_{j=0}^{l} \rho^{j-m} \langle |A_j| v, v\rangle , \end{aligned}$$

hence, $\langle |A|(\rho)v, v\rangle > \langle \rho^m v, v\rangle$, which contradicts Assumption 2.1.

In the case $l > m$, when $\rho \in (\rho_1, \rho_2)$ satisfies Assumption 2.1 and $\mathbb{S}(\rho, v) \neq \varnothing$ for some nonzero $v \in \mathcal{H}$, there can exist eigenvalues on $\mathbb{T}_\rho$ with corresponding Jordan chains of length more than one. This is shown in the following example.

Example 2.15. Let

$$A(\lambda) = \begin{bmatrix} \frac{2}{3}\lambda^3 + \frac{1}{3} & 0 \\ 0 & \frac{5}{12} \end{bmatrix}$$

and $F(\lambda) = \lambda^2 I_2 - A(\lambda) = \begin{bmatrix} -\frac{2}{3}\lambda^3+\lambda^2-\frac{1}{3} & 0 \\ 0 & \lambda^2-\frac{5}{12} \end{bmatrix}$.

Then we have that $|A| = A$ and $\phi_{|A|}(\tau) = \operatorname{spr} A(\tau) = \begin{cases} \frac{5}{12}, & \tau \leqslant \frac{1}{2} \\ \frac{2}{3}\tau^3 + \frac{1}{3}, & \tau > \frac{1}{2} \end{cases}$. Since

$$\phi_{|A|}(\rho = 1) = 1 = \rho^2,$$

$\rho = 1$ satisfies Assumption 2.1, see Lemma 2.4. Since $F(1) = \begin{bmatrix} 0 & 0 \\ 0 & \frac{7}{12} \end{bmatrix}$, we obtain that ρ is an eigenvalue of F and that the corresponding eigenspace is one dimensional. A corresponding eigenvector is given by $v = \left[\begin{smallmatrix} 1 \\ 0 \end{smallmatrix}\right]$. We have $F'(1) = \left[\begin{smallmatrix} 0 & 0 \\ 0 & 2 \end{smallmatrix}\right]$. Hence, the equation $F(1)w + F'(1)v = 0$ has a solution $w = \left[\begin{smallmatrix} 1 \\ 0 \end{smallmatrix}\right]$, so v, w constitute a Jordan chain of length 2 for F to the eigenvalue $\rho = 1$.

3 Degree reduction of m–monic matrix polynomials and preservation of spectral properties

The aim of this chapter is to develop degree reductions of a given m–monic matrix polynomial P to a 1-monic matrix polynomial $\mathcal{P}$.

We start with a general polynomial Q with coefficients that are elements of a complex algebra and introduce a reduction to degree $1 \leqslant k \leqslant l$. For $k = 1$ this will be e.g. the classical linearization via the companion form. In the second section of this chapter the correspondence between factorizations of the original polynomial Q and its reduction $\mathcal{Q}$ is developed. In both sections, the reduction for the m–monic case, which is a reduction by $m-1$, is formulated separately.

In the last section we focus on polynomials with matrix coefficients and investigate the recovery of Jordan chains of the original polynomial from the reduction.

3.1 Degree reduction of polynomials with coefficients that are elements of a complex Banach algebra

Consider a complex algebra $\mathfrak{A}$ with identity I. Let $\nu \in \mathbb{N}$ and denote by $\mathfrak{A}^{\nu,\nu}$ the set of $\nu \times \nu$ matrices with entries which are elements of $\mathfrak{A}$ equipped with the normal matrix multiplication induced by the multiplication in $\mathfrak{A}$. Clearly, $\mathfrak{A}^{\nu,\nu}$ is an algebra. We will also call $\mathcal{A} \in \mathfrak{A}^{\nu,\nu}$ a **block algebra matrix** of order ν.

Let Q be a polynomial with

$$Q(\lambda) = \sum_{j=0}^{l} \lambda^j Q_j$$

of degree $l \in \mathbb{N}$ where $Q_j \in \mathfrak{A}$ for all $j \in \langle l \rangle_0$. For simplicity we will call such a polynomial an **algebra polynomial**. Furthermore, let $\mathcal{Q}$ be the algebra polynomial with

$$\mathcal{Q}(\lambda) = \sum_{j=0}^{k} \lambda^j \mathcal{Q}_j$$

of degree $k \in \mathbb{N}$, $k < l$ with $\mathcal{Q}_j \in \mathfrak{A}^{m,m}$ for some $m \in \mathbb{N}$.

In view of unimodular matrix polynomials (i. e. matrix polynomials with constant nonzero determinant) we call an algebra polynomial Q **unimodular** if $Q(\lambda)$ is invertible for all $\lambda \in \mathbb{C}$. Note that if $\mathfrak{A} = \mathbb{C}^{n,n}$, this notion coincides with the notion of unimodular matrix polynomials. Indeed, since if the matrix $Q(\lambda)$ is invertible for all $\lambda \in \mathbb{C}$, the scalar polynomial $\det Q(\lambda)$ has no zeros and thus it has to be a nonzero constant.

For $\nu \in \mathbb{N}$ define

$$\mathcal{I}_\nu = \begin{bmatrix} I & & \\ & \ddots & \\ & & I \end{bmatrix} \in \mathfrak{A}^{\nu,\nu}.$$

We will call $\mathcal{Q}$ a **reduction** of Q to degree k with $1 \leqslant k \leqslant l$ if there are unimodular algebra polynomials $\mathcal{E}$ and $\mathcal{F}$ such that

$$\mathcal{E}(\lambda)\mathcal{Q}(\lambda)\mathcal{F}(\lambda) = \begin{bmatrix} Q(\lambda) & \\ & \mathcal{I}_{m-1} \end{bmatrix} =: Q(\lambda) \oplus \mathcal{I}_{m-1}$$

In analogy to the equivalence of matrix polynomials we say that $\mathcal{Q}$ and $Q \oplus \mathcal{I}_{m-1}$ are **equivalent**.

The next proposition was established through private communication with B.Nagy [Nag07].

Proposition 3.1. *Let Q be an algebra polynomial with $Q(\lambda) = \sum\limits_{j=0}^{l} \lambda^j Q_j$ and $Q_j \in \mathfrak{A}$ for all $j \in \langle l \rangle_0$. For $k \in \mathbb{N}$, $1 \leqslant k \leqslant l$ define the $(l-k+1) \times (l-k+1)$ block algebra matrices*

$$\mathcal{Q}_0 = \begin{bmatrix} Q_{l-k} & \cdots & \cdots & Q_0 \\ -I & & & \\ & \ddots & & \\ & & -I & 0 \end{bmatrix}, \quad \mathcal{Q}_1 = \begin{bmatrix} Q_{l-k-1} & & & \\ & I & & \\ & & \ddots & \\ & & & I \end{bmatrix},$$

$$\mathcal{Q}_j = \begin{bmatrix} Q_{l-k+j} & & & \\ & 0 & & \\ & & \ddots & \\ & & & 0 \end{bmatrix} \qquad \text{for } j = 2, \ldots, k$$

and define the algebra polynomial $\mathcal{Q}$ with

$$\mathcal{Q}(\lambda) = \sum_{j=0}^{k} \lambda^j \mathcal{Q}_j = \begin{bmatrix} \sum\limits_{j=0}^{k} \lambda^j Q_{l-k+j} & Q_{l-k-1} & \cdots & Q_0 \\ -I & \lambda I & & \\ & \ddots & \ddots & \\ & & -I & \lambda I \end{bmatrix}.$$

Then $\mathcal{Q}$ is a reduction of Q to degree k. Furthermore, we have

$$\mathcal{E}(\lambda)\mathcal{Q}(\lambda)\mathcal{F}(\lambda) = Q(\lambda) \oplus \mathcal{I}_{l-k}, \tag{3-27}$$

where $\mathcal{E}$ and $\mathcal{F}$ are algebra polynomials defined as follows:

$$\mathcal{E}(\lambda) = \begin{bmatrix} I & Q'_{l-k}(\lambda) & Q'_{l-k-1}(\lambda) & \cdots & Q'_1(\lambda) \\ & -I & -\lambda I & \cdots & -\lambda^{l-k-1} I \\ & & \ddots & \ddots & \vdots \\ & & & -I & -\lambda I \\ & & & & -I \end{bmatrix},$$

where Q'_j are matrix polynomials recursively defined by

$$Q'_0(\lambda) := Q(\lambda) \quad \text{and} \quad Q'_j(\lambda) := \frac{1}{\lambda}\left(Q'_{j-1}(\lambda) - Q_{j-1}\right), \; j \in \langle l-k \rangle.$$

$$\mathcal{F}(\lambda) = \begin{bmatrix} \lambda^{l-k} I & I & & & \\ \lambda^{l-k-1} I & 0 & \ddots & & \\ \vdots & & \ddots & \ddots & \\ \lambda I & & & \ddots & I \\ I & & & & 0 \end{bmatrix}$$

Proof. We verify (3-27) by direct calculation. First we have

$$\mathcal{Q}(\lambda)F(\lambda) = \begin{bmatrix} Q(\lambda) & \sum_{j=0}^{k} \lambda^j Q_{l-k+j} & Q_{l-k-1} & \cdots & Q_1 \\ & -I & \lambda I & & \\ & & \ddots & \ddots & \\ & & & \ddots & \lambda I \\ & & & & -I \end{bmatrix}$$

$$= \begin{bmatrix} Q(\lambda) & Q'_{l-k}(\lambda) & Q_{l-k-1} & \cdots & Q_1 \\ & -I & \lambda I & & \\ & & \ddots & \ddots & \\ & & & \ddots & \lambda I \\ & & & & -I \end{bmatrix}.$$

Hence, the first row of $\mathcal{E}(\lambda)\mathcal{Q}(\lambda)\mathcal{F}(\lambda)$ is $\begin{bmatrix} Q(\lambda) \\ 0 \\ Q_{l-k-1}+\lambda Q'_{l-k}(\lambda)-Q'_{l-k-1}(\lambda) \\ Q_{l-k-2}+\lambda Q'_{l-k-1}(\lambda)-Q'_{l-k-2}(\lambda) \\ \vdots \\ Q_1+\lambda Q'_2(\lambda)-Q'_1(\lambda) \end{bmatrix}^t$,

where for $B_1, \ldots, B_k \in \mathfrak{A}$ we define $\begin{bmatrix} B_1 \\ \vdots \\ B_k \end{bmatrix}^t = \begin{bmatrix} B_1 & \cdots & B_k \end{bmatrix}$.

Note that for $p \in \langle\, l-k-1 \,\rangle$, by definition of the polynomials Q'_j,

$$\begin{aligned} &Q_{l-k-p} + \lambda Q'_{l-k-p+1}(\lambda) - Q'_{l-k-p}(\lambda) \\ &= Q_{l-k-p} + Q'_{l-k-p}(\lambda) - Q_{l-k-p} - Q'_{l-k-p}(\lambda) = 0, \end{aligned}$$

hence

$$\mathcal{E}(\lambda)\mathcal{Q}(\lambda)\mathcal{F}(\lambda) = \begin{bmatrix} Q(\lambda) & & & \\ & I & & \\ & & \ddots & \\ & & & I \end{bmatrix}.$$

□

Remark 3.2. *Since $\mathcal{E}(\lambda)$ and $\mathcal{F}(\lambda)$ are invertible for all $\lambda \in \mathbb{C}$, it is immediately clear that $Q(\lambda)$ is invertible if and only if $\mathcal{Q}(\lambda)$ is invertible.*

We will call the reduction $\mathcal{Q}$ as depicted in Proposition 3.1 the **canonical reduction** of the algebra polynomial Q to degree k and will consequently denote this reduction and its coefficients with calligraphic letters. Note that the canonical reduction to degree $k = l$ leaves the original polynomial unchanged and for $k = 1$ the canonical reduction coincides with the linearization via a companion form.

If we consider an m–monic algebra polynomial $P(\lambda) = \lambda^m I - \sum_{j=0}^{l} \lambda^j A_j$ we can apply Proposition 3.1 with $Q_j = \delta_{mj} I - A_j$ and $k = l - m + 1$ and immediately obtain the canonical reduction $\mathcal{P}$ of P. We formulate this in the following corollary.

Corollary 3.3. *Let P be an m–monic algebra polynomial, with*

$$P(\lambda) = \lambda^m I - A(\lambda) = \lambda^m I - (\lambda^l A_l + \cdots - \lambda A_1 + A_0), \tag{3-28}$$

and $A_j \in \mathfrak{A}$ for all $j \in \langle l \rangle$. Define the $m \times m$ algebra polynomial

$$\mathcal{P}(\lambda) = \lambda \mathcal{I}_m - \mathcal{A}(\lambda) = \lambda \mathcal{I}_m - (\lambda^{l-m+1} \mathcal{A}_{l-m+1} + \cdots + \lambda \mathcal{A}_1 + \mathcal{A}_0), \tag{3-29}$$

where

$$\mathcal{A}_0 = \begin{bmatrix} A_{m-1} & \cdots & \cdots & A_0 \\ I & & & \vdots \\ & \ddots & & \vdots \\ & & I & 0 \end{bmatrix} \tag{3-30}$$

and

$$\mathcal{A}_j = \begin{bmatrix} A_{j+m-1} & & & \\ & 0 & & \\ & & \ddots & \\ & & & 0 \end{bmatrix} \quad \text{for } j = 1, \ldots, l - m + 1. \tag{3-31}$$

Then $\mathcal{P}$ is equivalent to $P(\lambda) \oplus \mathcal{I}_{m-1}$ via the unimodular transformation

$$\mathcal{E}(\lambda)\mathcal{P}(\lambda)\mathcal{F}(\lambda) = P(\lambda) \oplus \mathcal{I}_{m-1} = \begin{bmatrix} P(\lambda) & \\ & \mathcal{I}_{m-1} \end{bmatrix},$$

where $\mathcal{E}(\lambda)$ and $\mathcal{F}(\lambda)$ are defined as follows.

$$\mathcal{E}(\lambda) = \begin{bmatrix} I & P'_{m-1}(\lambda) & P'_{m-2}(\lambda) & \cdots & P'_1(\lambda) \\ & -I & -\lambda I & \cdots & -\lambda^{m-2} I \\ & & \ddots & \ddots & \vdots \\ & & & -I & -\lambda I \\ & & & & -I \end{bmatrix},$$

and $P'_j(\lambda)$ are algebra polynomials recursively defined by

$$P'_0(\lambda) := P(\lambda) \quad \text{and} \quad P'_{j+1}(\lambda) := \frac{1}{\lambda}(P'_j(\lambda) + A_j), \; j \in \langle m - 2 \rangle .$$

The right transformation is given by

$$\mathcal{F}(\lambda) = \begin{bmatrix} \lambda^{m-1} I & I & & & \\ \lambda^{m-2} I & 0 & \ddots & & \\ \vdots & & \ddots & \ddots & \\ \lambda I & & & \ddots & I \\ I & & & & 0 \end{bmatrix}.$$

3.2 Reduction and factorizations

In this section we will point out the connection between certain factorizations of the original polynomial Q and factorizations of its canonical reduction $\mathcal{Q}$ to degree $k < l$ in the sense of Proposition 3.1. Clearly, nothing has to be investigated if $k = l$. We will see that if $\mathcal{Q}$ has a factorization of the type

$$\mathcal{Q}(\lambda) = \mathcal{B}(\lambda)(\lambda\mathcal{I}_{l-k+1} - \mathcal{C}),$$

then $\mathcal{C}$ has to be a certain block algebra companion matrix (for simplicity, in the following, we will simply talk of companion matrices). This is due to the special structure of the canonical reduction introduced in Section 3.1.

Theorem 3.4. *Consider the algebra polynomial* $Q(\lambda) = \sum_{j=0}^{l} \lambda^j Q_j$ *and its reduction* $\mathcal{Q}(\lambda) = \sum_{j=0}^{k} \lambda^j \mathcal{Q}_j$ *to degree* $k < l$ *according to Proposition 3.1. Then the following assertions hold.*

(i) *If Q has a factorization*

$$Q(\lambda) = B(\lambda)C(\lambda) = \left(\sum_{j=0}^{k-1} \lambda^j B_j\right)\left(\lambda^{l-k+1} I - \sum_{j=0}^{l-k} \lambda^j C_j\right), \tag{3-32}$$

then $\mathcal{Q}$ has a factorization

$$\mathcal{Q}(\lambda) = \mathcal{B}(\lambda)(\lambda\mathcal{I}_{l-k+1} - \mathcal{C}) \tag{3-33}$$

where $\mathcal{C} = \begin{bmatrix} C_{l-k} & \cdots & \cdots & C_0 \\ I & & & \\ & \ddots & & \\ & & I & 0 \end{bmatrix} \in \mathfrak{A}^{l-k+1,l-k+1}$ *is the companion matrix of C.*

(ii) *If $\mathcal{Q}$ has the factorization* (3-33), *then the block algebra matrix $\mathcal{C}$ in the right factor is of the form* $\mathcal{C} = \begin{bmatrix} * & * & * & * \\ I & & & \\ & \ddots & & \\ & & I & 0 \end{bmatrix}$, *and Q has the factorization* (3-32), *such that $\mathcal{C}$ is precisely the companion matrix of C.*

Proof. Set $m = l - k + 1$. Suppose that (3-32) holds. This is equivalent to the system of equations

$$Q_j = B_{j-m} - \sum_{i=1}^{m} B_{j-m+i} C_{m-i} \quad (j \in \langle l \rangle_0), \tag{3-34}$$

where we set $B_\nu = 0$ if $\nu \notin \langle l - m \rangle_0$. Recursively define the block algebra matrices $\mathcal{B}_0, \ldots, \mathcal{B}_{l-m}$ by

$$\mathcal{B}_{l-m} = \mathcal{Q}_{l-m+1}, \tag{3-35}$$

$$\mathcal{B}_j = \mathcal{Q}_{j+1} + \mathcal{B}_{j+1}\mathcal{C} \quad (j = l - m - 1, \ldots, 0). \tag{3-36}$$

Then $\mathcal{Q}$ has the factorization $\mathcal{Q}(\lambda) = \left(\sum_{j=0}^{l-m} \lambda^j \mathcal{B}_j\right)(\lambda\mathcal{I}_{l-k+1} - \mathcal{C})$ if only if the remaining equation

$$\mathcal{Q}_0 = -\mathcal{B}_0\mathcal{C}, \tag{3-37}$$

holds, which is arising from comparing coefficients on both sides of the factorization.

In what follows, it is helpful to simplify the notation.

For $v = [\, v_1 \cdots v_m \,]^T \in \mathbb{C}^m$, a complex algebra $\mathfrak{A}$ and $T \in \mathfrak{A}$, we write

$$v \otimes T := \begin{bmatrix} v_1 T \\ \vdots \\ v_m T \end{bmatrix}.$$

If $\mathfrak{B} = \mathbb{C}^{n,n}$, $v \otimes T$ is precisely the Kronecker product of v with T. It is easily checked that basic algebraic operations of the standard Kronecker product (see e.g. [HJ91]) transfer to our more general case.

In what follows, the identities

$$\begin{aligned} \mathcal{B}_{l-m-\nu} = e_1 \otimes \Bigg[\; B_{l-m-\nu} \;\Bigg|\; \sum_{i=2}^{m} B_{l-m-\nu+i-1} C_{m-i} \;\Bigg|\; \sum_{i=3}^{m} B_{l-m-\nu+i-2} C_{m-i} \;\Bigg|\; \cdots \\ \cdots \;\Bigg|\; \sum_{i=m-1}^{m} B_{l-m-\nu+i-(m-2)} C_{m-i} \;\Bigg|\; B_{l-m-\nu+1} C_0 \;\Bigg] + \delta_{\nu,l-m} \begin{bmatrix} 0 & I & & \\ & & \ddots & \\ & & & I \end{bmatrix}, \end{aligned} \tag{3-38}$$

for $\nu = 0, \ldots, l-m$ will be helpful, where $e_1 \in \mathbb{C}^m$ denotes the first unit vector, δ_{ij} the Kronecker symbol and again $B_j = 0$ if $j \notin \langle\, l-m \,\rangle_0$. The identities (3-38) are easily, even though not without some technical effort, verified by induction. First, due to (3-35) and (3-34) we have

$$\mathcal{B}_{l-m} = \mathcal{Q}_{l-m+1} = \begin{bmatrix} Q_l & & \\ & 0 & \\ & & \ddots \\ & & & 0 \end{bmatrix} = e_1 \otimes \begin{bmatrix} B_{l-m} & 0 & \cdots & 0 \end{bmatrix}.$$

Suppose now that (3-38) holds for some $\nu \in \langle\, l-m-2 \,\rangle_0$. Then by (3-36)

$$\mathcal{B}_{l-m-(\nu+1)} = \mathcal{Q}_{l-m-\nu} + \mathcal{B}_{l-m-\nu}\mathcal{C}, \tag{3-39}$$

where

$$\begin{aligned} \mathcal{B}_{l-m-\nu}\mathcal{C} &= e_1 \otimes \Bigg(\Bigg[\; B_{l-m-\nu} \;\Bigg|\; \sum_{i=2}^{m} B_{l-m-\nu+i-1} C_{m-i} \;\Bigg|\; \sum_{i=3}^{m} B_{l-m-\nu+i-2} C_{m-i} \;\Bigg|\; \cdots \\ &\qquad \cdots \;\Bigg|\; \sum_{i=m-1}^{m} B_{l-m-\nu+i-(m-2)} C_{m-i} \;\Bigg|\; B_{l-m-\nu+1} C_0 \;\Bigg]\Bigg)\mathcal{C} \\ &= e_1 \otimes \Bigg[\; B_{l-m-\nu} C_{m-1} + \sum_{i=2}^{m} B_{l-m-\nu+i-1} C_{m-i} \;\Bigg|\; B_{l-m-\nu} C_{m-2} \\ &\qquad + \sum_{i=3}^{m} B_{l-m-\nu+i-2} C_{m-i} \;\Bigg|\; \cdots \;\Bigg|\; B_{l-m-\nu} C_1 + B_{l-m-\nu+1} C_0 \;\Bigg|\; B_{l-m-\nu} C_0 \;\Bigg] \\ &= e_1 \otimes \Bigg[\; \sum_{i=1}^{m} B_{l-m-\nu+i-1} C_{m-i} \;\Bigg|\; \sum_{i=2}^{m} B_{l-m-\nu+i-2} C_{m-i} \;\Bigg|\; \cdots \\ &\qquad \cdots \;\Bigg|\; \sum_{i=m-1}^{m} B_{l-m-\nu+i-(m-1)} C_{m-i} \;\Bigg|\; B_{l-m-\nu} C_0 \;\Bigg]. \end{aligned}$$

Since $\mathcal{Q}_{l-m-\nu} = \begin{bmatrix} Q_{l-\nu-1} & & & \\ & 0 & & \\ & & \ddots & \\ & & & 0 \end{bmatrix}$, we have

$$\mathcal{B}_{l-m-(\nu+1)} = e_1 \otimes \Big[\; B_{l-m-\nu-1} \;\Big|\; \sum_{i=2}^{m} B_{l-m-(\nu+1)+i-1} C_{m-i} \;\Big|\; \cdots$$
$$\cdots \;\Big|\; \sum_{i=m-1}^{m} B_{l-m-(\nu+1)+i-(m-2)} C_{m-i} \;\Big|\; B_{l-m-(\nu+1)+1} C_0 \; \Big],$$

thus (3-38) holds for all $\nu \in \langle\, l-m-1 \,\rangle_0$. The induction step from $\nu = l-m-1$ to $\nu = l-m$ works quite similar. In this step, equation (3-39) takes the form $\mathcal{B}_0 = \mathcal{Q}_1 + \mathcal{B}_1 \mathcal{C}$.

Therefore, the summand $\delta_{\nu,l-m} \begin{bmatrix} 0 & & & \\ & I & & \\ & & \ddots & \\ & & & I \end{bmatrix}$ appears in (3-38), since $\mathcal{Q}_1$ has a structure different from $\mathcal{Q}_2, \ldots, \mathcal{Q}_{l-m+1}$, more precisely $Q_1 = \begin{bmatrix} Q_m & & & \\ & I & & \\ & & \ddots & \\ & & & I \end{bmatrix}$.

Thus (3-38) holds for all $\nu \in \langle\, l-m \,\rangle_0$.

We are now able to check the remaining identity (3-37). Indeed by (3-38) we have

$$\mathcal{B}_0 = \begin{bmatrix} B_0 & \sum_{i=2}^{m} B_{i-1} C_{m-i} & \sum_{i=3}^{m} B_{i-2} C_{m-i} & \cdots & \sum_{i=m-1}^{m} B_{i-(m-2)} C_{m-i} & B_1 C_0 \\ & I & & & & \\ & & I & & & \\ & & & I & & \\ & & & & \ddots & \\ & & & & & I \end{bmatrix} \tag{3-40}$$

and therefore

$$-\mathcal{B}_0 \mathcal{C} = e_1 \otimes \Big[\; -B_0 C_{m-1} - \sum_{i=2}^{m} B_{i-1} C_{m-i} \;\Big|\; -B_0 C_{m-2} - \sum_{i=3}^{m} B_{i-2} C_{m-i} \;\Big|\; \cdots$$
$$\cdots \;\Big|\; -B_0 C_1 - B_1 C_0 \;\Big|\; -B_0 C_0 \; \Big] - \begin{bmatrix} 0 & & & \\ I & \ddots & & \\ & \ddots & \ddots & \\ & & I & 0 \end{bmatrix}$$
$$= e_1 \otimes \Big[\; -\sum_{i=1}^{m} B_{i-1} C_{m-i} \;\Big|\; -\sum_{i=2}^{m} B_{i-2} C_{m-i} \;\Big|\; \cdots$$
$$\cdots \;\Big|\; -\sum_{i=m-1}^{m} B_{i-(m-1)} C_{m-i} \;\Big|\; -B_0 C_0 \; \Big] - \begin{bmatrix} 0 & & & \\ I & \ddots & & \\ & \ddots & \ddots & \\ & & I & 0 \end{bmatrix}$$
$$= \begin{bmatrix} Q_{m-1} & Q_{m-2} & \cdots & Q_0 \\ -I & \ddots & & \\ & \ddots & \ddots & \\ & & -I & 0 \end{bmatrix} = \mathcal{Q}_0.$$

This proves part (i) of the theorem.

Suppose now that $\mathcal{C} \in \mathfrak{A}^{l-k+1,l-k+1}$ is such that

$$\mathcal{Q}(\lambda) = \left(\sum_{j=0}^{l-m} \lambda^j \mathcal{B}_j\right)(\lambda \mathcal{I}_{l-k+1} - \mathcal{C}) \tag{3-41}$$

with some $m \times m$ block algebra matrices $\mathcal{B}_0, \ldots, \mathcal{B}_{l-m}$. Therefore, the equations (3-35), (3-36) and (3-37) are valid:

$$\begin{aligned} \mathcal{B}_{l-m} &= \mathcal{Q}_{l-m+1}, \\ \mathcal{B}_j &= \mathcal{Q}_{j+1} + \mathcal{B}_{j+1}\mathcal{C} \quad (j = l-m-1, \ldots, 0), \\ \mathcal{Q}_0 &= -\mathcal{B}_0\mathcal{C}. \end{aligned} \tag{3-42}$$

They imply that for $j = l-m-1, \ldots, 1$ the matrices $\mathcal{B}_j$ are of the form

$$\mathcal{B}_j = \begin{bmatrix} * & * & * & * \\ & 0 & & \\ & & \ddots & \\ & & & 0 \end{bmatrix} \quad \text{and} \quad \mathcal{B}_0 = \mathcal{Q}_1 + \mathcal{B}_1\mathcal{C} = \begin{bmatrix} * & * & * & * \\ & I & & \\ & & \ddots & \\ & & & I \end{bmatrix}. \tag{3-43}$$

Finally from $\begin{bmatrix} Q_{m-1} & \cdots & \cdots & Q_0 \\ -I & & & \\ & \ddots & & \\ & & -I & 0 \end{bmatrix} = \mathcal{Q}_0 = -\mathcal{B}_0\mathcal{C}$, it follows that $\mathcal{C}$ has the form

$$\mathcal{C} = \begin{bmatrix} * & * & * & * \\ I & & & \\ & \ddots & & \\ & & I & 0 \end{bmatrix},$$

i. e., $\mathcal{C}$ is a companion matrix. In order to finish the proof of (ii), for $j \in \langle m-1 \rangle_0$ one has to define $C_j = \mathcal{C}_{1(m-j)}$, where $\mathcal{C}_{\mu\nu}$ denotes the element in the μ-th block row and ν-th block column of $\mathcal{C}$ (If the block algebra matrix already has a subscript, e.g. $\mathcal{C}_j$, we also write $(\mathcal{C}_j)_{\mu\nu}$). Furthermore, for $j \in \langle l-m \rangle_0$ define B_j to be the element in the upper left algebra block of $\mathcal{B}_j$.

It remains to show that (3-34) holds. At first suppose that $j \in \langle m-1 \rangle_0$. Then

$$Q_j = (\mathcal{Q}_0)_{1(m-j)} = -(\mathcal{B}_0\mathcal{C})_{1(m-j)} = -B_0C_j - (\mathcal{B}_0)_{1(m-j+1)}.$$

Using (3-42) we can calculate $(\mathcal{B}_0)_{1(m-j+1)}$ by

$$(\mathcal{B}_0)_{1(m-j+1)} = (\mathcal{Q}_1 - \mathcal{B}_1\mathcal{C})_{1(m-j+1)} = (\mathcal{B}_1\mathcal{C})_{1(m-j+1)} = B_1C_{j-1} + (\mathcal{B}_1)_{1(m-j+2)}.$$

Again by (3-42)

$$(\mathcal{B}_1)_{1(m-j+2)} = (\mathcal{Q}_2 - \mathcal{B}_2\mathcal{C})_{1(m-j+2)} = (\mathcal{B}_2\mathcal{C})_{1(m-j+2)} = B_2C_{j-2} + (\mathcal{B}_2)_{1(m-j+3)}.$$

This procedure can be repeated until

$$(\mathcal{B}_{j-2})_{1(m-1)} = (\mathcal{Q}_{j-1} + \mathcal{B}_{j-1}\mathcal{C})_{1(m-1)} = (\mathcal{B}_{j-1}\mathcal{C})_{1(m-1)} = B_{j-1}C_1 + (\mathcal{B}_{j-1})_{1m},$$

where

$$(\mathcal{B}_{j-1})_{1m} = (\mathcal{Q}_j)_{1m} + (\mathcal{B}_j\mathcal{C})_{1m} = (\mathcal{B}_j\mathcal{C})_{1m} = B_jC_0.$$

Again, we set $B_\nu = 0$ when $\nu \notin \langle l-m \rangle_0$. Now, putting things together we obtain

$$Q_j = -\sum_{i=0}^{j} B_{j-i}C_i = -\sum_{i=0}^{m-1} B_{j-i}C_i = -\sum_{i=1}^{m} B_{j-m+i}C_{m-i} \quad \text{for} \quad j \in \langle m-1 \rangle_0.$$

This is precisely (3-34) for $j \in \langle m-1 \rangle_0$.

Suppose now that $j \in \{m, \dots, l\}$. Then

$$\begin{aligned} Q_j &= (\mathcal{Q}_{j-m+1})_{11} = (\mathcal{B}_{j-m})_{11} - (\mathcal{B}_{j-m+1}\mathcal{C})_{11} \\ &= B_{j-m} - B_{j-m+1}C_{m-1} - (\mathcal{B}_{j-m+1})_{12}. \end{aligned}$$

Because of $(\mathcal{Q}_j)_{1\nu} = 0$ for $\nu \neq 1$, we have

$$(\mathcal{B}_{j-m+1})_{12} = (\mathcal{B}_{j-m+2}\mathcal{C})_{12} = B_{j-m+2}C_{m-2} + (\mathcal{B}_{j-m+2})_{13},$$

$$(\mathcal{B}_{j-m+2})_{13} = (\mathcal{B}_{j-m+3}\mathcal{C})_{13} = B_{j-m+3}C_{m-3} + (\mathcal{B}_{j-m+3})_{14}.$$

Repeating this until

$$(\mathcal{B}_{j-2})_{1(m-1)} = (\mathcal{B}_{j-1}\mathcal{C})_{1(m-1)} = B_{j-1}C_1 + (\mathcal{B}_{j-1})_{1m},$$

$$(\mathcal{B}_{j-1})_{1m} = B_jC_0,$$

we get

$$Q_j = B_{j-m} - \sum_{i=1}^{m} B_{j-m+i}C_{m-i} \quad \text{for} \quad j = m, \dots, l.$$

Hence, (3-34) holds for all $j \in \langle l \rangle_0$ and part (*ii*) of the theorem is proved. □

From the proof of Theorem 3.4 it also follows how to obtain the coefficients of the left factors of one factorization from the other one. For clarity, we will repeat these identities in the following remark.

Remark 3.5. (*i*) *Suppose that Q has a factorization* (3-32). *Then the coefficients of the left factor of the factorization* (3-33) *of $\mathcal{Q}$ can be obtained by identity* (3-38):

$$\mathcal{B}_{l-m-\nu} = e_1 \otimes \left[\; B_{l-m-\nu} \;\middle|\; \sum_{i=2}^{m} B_{l-m-\nu+i-1}C_{m-i} \;\middle|\; \sum_{i=3}^{m} B_{l-m-\nu+i-2}C_{m-i} \;\middle|\; \cdots \right.$$
$$\left. \cdots \;\middle|\; \sum_{i=m-1}^{m} B_{l-m-\nu+i-(m-2)}C_{m-i} \;\middle|\; B_{l-m-\nu+1}C_0 \; \right] + \delta_{\nu,l-m} \begin{bmatrix} 0 & I & & \\ & & \ddots & \\ & & & I \end{bmatrix},$$

for $\nu = 0, \dots, l-m$.

(*ii*) *Suppose that $\mathcal{Q}$ has the factorization* (3-33). *Then the coefficients B_j of the left factor of the factorization* (3-32) *can be obtained by*

$$B_j = (\mathcal{B}_j)_{11} \quad \textit{for} \quad j \in \langle l-m \rangle_0,$$

where $(\mathcal{B}_j)_{11}$ denotes the upper left algebra block of $\mathcal{B}_j$.

We reformulate Theorem 3.4 for the m–monic case in the following corollary.

Corollary 3.6. *Let $m < l$ and consider the m–monic algebra polynomial P with $P(\lambda) = \lambda^m I - \sum_{j=0}^{l} \lambda^j A_j$ and its canonical reduction $\mathcal{P}$. Then the following assertions hold.*

(*i*) *If P has a factorization*

$$P(\lambda) = B(\lambda)C(\lambda) = \left(\sum_{j=0}^{l-m} \lambda^j B_j\right)\left(\lambda^m I - \sum_{j=0}^{m-1} \lambda^j C_j\right), \tag{3-44}$$

then $\mathcal{P}$ has a right divisor $\lambda\mathcal{I}_m - \mathcal{C}$, where

$$\mathcal{C} = \begin{bmatrix} C_{m-1} & \cdots & \cdots & C_0 \\ I & & & \\ & \ddots & & \\ & & I & 0 \end{bmatrix}$$

is the companion matrix of C.

(*ii*) *If $\lambda\mathcal{I}_m - \mathcal{C}$ is any right divisor of $\mathcal{P}$, then $\mathcal{C}$ is of the form $\mathcal{C} = \begin{bmatrix} * & * & * & * \\ I & & & \\ & \ddots & & \\ & & I & 0 \end{bmatrix}$, and P has the factorization (3-44) such that $\mathcal{C}$ is the companion matrix of C.*

We state one more consequence for the case that the algebra $\mathfrak{A}$ is the space of $n \times n$ matrices. It follows from the Perron-Frobenius theorem and the fact that $\mathcal{C}$ is the companion matrix of the right factor $C(\lambda)$ in (3-44), which means that C and $\mathcal{C}$ have the same eigenvalues.

Corollary 3.7. *Suppose that $\mathfrak{A} = \mathbb{C}^{n,n}$, i. e., P is an m–monic $n \times n$ matrix polynomial and the right factor $C(\lambda)$ in the factorization* (3-44) *has coefficients C_j that are entrywise nonnegative matrices for all $j \in \langle m-1 \rangle_0$. Then the spectral radius* $\operatorname{spr}(C)$ *of C is an eigenvalue of C and there is an entrywise nonnegative eigenvector corresponding to* $\operatorname{spr}(C)$.

3.3 Recovery of Jordan chains

We now concentrate on the case when the coefficients of the polynomial Q are in the complex Banach algebra $\mathbb{C}^{n,n}$ of complex $n \times n$-matrices. The connection between the eigenvalues, eigenvectors and Jordan chains of Q and $\mathcal{Q}$ are investigated in this section.

From Remark 3.2 it follows that a canonical reduction $\mathcal{Q}$ of a matrix polynomial Q has the same finite eigenvalues as Q. Furthermore, due to the equivalence of Q and $\mathcal{Q}$, they have the same Smith form and therewith the lengths of the Jordan chains of Q and $\mathcal{Q}$ associated with finite eigenvalues coincide. See e.g. [Rod89], [GLR82].

Define the operator valued function

$$\Lambda : \mathbb{C} \to L(\mathbb{C}^n, \mathbb{C}^{(l-k+1)n}), \quad \Lambda(\lambda)u = \begin{bmatrix} \lambda^{l-k} \\ \vdots \\ \lambda \\ 1 \end{bmatrix} \otimes u, \quad \text{for } u \in \mathbb{C}^n. \tag{3-45}$$

In the following we will identify $\Lambda(\lambda) \in L(\mathbb{C}^n, \mathbb{C}^{(l-k+1)n})$ with its associated matrix $\begin{bmatrix}\lambda^{l-k} & \cdots & \lambda & 1\end{bmatrix}^T \otimes I_n$ with respect to the standard basis.

Denote by $\mathcal{Q}\Lambda : \mathbb{C} \to L(\mathbb{C}^n, \mathbb{C}^{(l-k+1)n})$ the operator valued function mapping $\lambda \in \mathbb{C}$ to $\mathcal{Q}(\lambda)\Lambda(\lambda)$.

The following lemma makes some technical preparations for the main result of this section.

Lemma 3.8. *Let Q and $\mathcal{Q}$ be as in Proposition 3.1 and Λ as in* (3-45). *Let $e_1 \in \mathbb{C}^{l-k+1}$ denote the first vector of the standard basis. Then*

(*i*) $(\mathcal{Q}\Lambda)(\lambda) = e_1 \otimes Q(\lambda)$ *for all $\lambda \in \mathbb{C}$;*

(*ii*) *Let $(x_\nu)_{\nu=0}^{\kappa-1} \subset \mathbb{C}^n$ be a sequence of vectors and let $\lambda_0 \in \mathbb{C}$. Define the sequence $(\xi_\nu)_{\nu=0}^{\kappa-1} \in \mathbb{C}^{(l-k+1)n}$ via*

$$\xi_\nu = \sum_{i=0}^{\nu} \frac{1}{i!} \Lambda^{(i)}(\lambda_0) x_{\nu-i}.$$

Then the identity

$$\sum_{j=0}^{\nu} \frac{1}{j!} \mathcal{Q}^{(j)}(\lambda_0)\xi_{\nu-j} = e_1 \otimes \sum_{j=0}^{\nu} \frac{1}{j!} Q^{(j)}(\lambda_0) x_{\nu-j}$$

holds for all $\nu \in \langle\, \kappa - 1\,\rangle_0$.

Proof. (*i*) This follows immediately from computing $\mathcal{Q}(\lambda)\Lambda(\lambda)$.

(*ii*) We verify this assertion by direct calculation.

$$\begin{aligned}
\sum_{j=0}^{\nu} \frac{1}{j!} \mathcal{Q}^{(j)}(\lambda_0)\xi_{\nu-j} &= \sum_{j=0}^{\nu} \frac{1}{j!} \mathcal{Q}^{(j)}(\lambda_0) \left(\sum_{i=0}^{\nu-j} \frac{1}{i!} \Lambda^{(i)}(\lambda_0) x_{\nu-j-i} \right) \\
&= \sum_{j=0}^{\nu} \frac{1}{j!} \mathcal{Q}^{(j)}(\lambda_0) \left(\sum_{i=0}^{\nu-j} \frac{1}{(\nu-j-i)!} \Lambda^{(\nu-j-i)}(\lambda_0) x_i \right) \\
&= \sum_{j=0}^{\nu} \sum_{i=0}^{\nu-j} \frac{1}{j!(\nu-j-i)!} \mathcal{Q}^{(j)}(\lambda_0) \Lambda^{(\nu-j-i)}(\lambda_0) x_i \\
&= \sum_{i=0}^{\nu} \sum_{j=0}^{\nu-i} \frac{1}{j!(\nu-i-j)!} \mathcal{Q}^{(j)}(\lambda_0) \Lambda^{(\nu-i-j)}(\lambda_0) x_i \\
&= \sum_{i=0}^{\nu} \frac{1}{(\nu-i)!} \left(\sum_{j=0}^{\nu-i} \binom{\nu-i}{j} \mathcal{Q}^{(j)}(\lambda_0) \Lambda^{(\nu-i-j)}(\lambda_0) \right) x_i \\
&= \sum_{i=0}^{\nu} \frac{1}{(\nu-i)!} (\mathcal{Q}\Lambda)^{(\nu-i)}(\lambda_0) x_i = \sum_{j=0}^{\nu} \frac{1}{j!} (\mathcal{Q}\Lambda)^{(j)}(\lambda_0) x_{\nu-j} \\
&= \sum_{j=0}^{\nu} \frac{1}{j!} \left(e_1 \otimes Q^{(j)} \right)(\lambda_0) x_{\nu-j} = e_1 \otimes \sum_{j=0}^{\nu} \frac{1}{j!} Q^{(j)}(\lambda_0) x_{\nu-j}.
\end{aligned}$$

□

Proposition 3.9. *Let Q and $\mathcal{Q}$ be as in Proposition 3.1 and let $\Lambda(\lambda) \in L(\mathbb{C}^n, \mathbb{C}^{(l-k+1)n})$ be defined as in (3-45). Then for $\kappa \in \mathbb{N} \setminus \{0\}$ the following assertions hold.*

(i) *$(x_\nu)_{\nu=0}^{\kappa-1}$ is a Jordan chain for Q to the eigenvalue $\lambda \in \mathbb{C}$ if and only if $(\xi_\nu)_{\nu=0}^{\kappa-1}$ with $\xi_\nu = \sum_{i=0}^{\nu} \frac{1}{i!}\Lambda^{(i)}(\lambda)x_{\nu-i}$ is a Jordan chain for $\mathcal{Q}$ to the eigenvalue λ.*

(ii) *If Q is regular and $\xi \in \mathbb{C}^{(l-k+1)n}$ is an eigenvector of $\mathcal{Q}$ to the eigenvalue λ, then $\xi = \Lambda(\lambda)x$, where $x \in \mathbb{C}^n$ is an eigenvector of Q to the eigenvalue λ.*

Proof. (*i*) From Lemma 3.8(*ii*) it immediately follows that $\sum_{j=0}^{\nu} \frac{1}{j!}\mathcal{Q}^{(j)}(\lambda_0)\xi_{\nu-j} = 0$ if and only if $\sum_{j=0}^{\nu} \frac{1}{j!}Q^{(j)}(\lambda_0)x_{\nu-j} = 0$.

(*ii*) The proof works in the same way as the proof of [MMMM06, Theorem 3.8].

□

The next corollary is a consequence of Remark 1.23.

Corollary 3.10. *A regular matrix polynomial Q has ln (finite and infinite) eigenvalues counting multiplicities. Since Q and its canonical reduction $\mathcal{Q}$ are equivalent via the unimodular matrix polynomials $\mathcal{E}$ and $\mathcal{F}$ from Lemma 3.3, it follows that $\mathcal{Q}$ is regular as well. It has $k(l-k+1)n$ (finite and infinite) eigenvalues counting multiplicities. Hence, the difference of these two numbers is*

$$k(l-k+1)n - ln = (l-k)(k-1)n \geqslant 0,$$

i. e., in general, $\mathcal{Q}$ has more eigenvalues than Q. Again, from the equivalence of Q and $\mathcal{Q}$ it follows that the finite *eigenvalues and partial multiplicities of Q and $\mathcal{Q}$ are identical. So the $(l-k)(k-1)n$ additional eigenvalues of $\mathcal{Q}$ are infinite eigenvalues.*

Note that we do not state anything about the partial multiplicities of the eigenvalue infinity. We only state that the algebraic multiplicity of infinity as an eigenvalue of Q does increase if we perform a canonical reduction. In general, partial multiplicities of the eigenvalue infinity are not faithfully recovered by degree reductions. For instance, suppose that infinity has the algebraic multiplicity $a(Q,\infty)$ as an eigenvalue of Q. Then for any combination of partial multiplicities compatible with $a(Q,\infty)$, one can find a linearization of Q (i. e. a linear matrix polynomial $\mathcal{L}$ and unimodular matrix polynomials $\mathcal{E}$ and $\mathcal{F}$ such that $\mathcal{E}(\lambda)\mathcal{L}(\lambda)\mathcal{F}(\lambda) = \begin{bmatrix} Q(\lambda) & \\ & I_{(l-1)n} \end{bmatrix}$) which realizes this structure, see [LP05]. To adjust this, special linearizations were introduced in [GKL88] and were called *strong linearizations* later in [LP05]. However, the problem of the conservation and change of the Jordan structure of Q at infinity after applying a canonical reduction, as we introduced in Section 3.1, is not studied in this thesis.

4 m–monic matrix polynomials with entrywise nonnegative coefficients

In this chapter we will deal with m–monic matrix polynomials

$$P(\lambda) = \lambda^m I_n - A(\lambda),$$

where A is a matrix polynomial of degree $l > m$ with entrywise nonnegative coefficients. In [PT04], matrix polynomials of the form $L(\lambda) = \lambda^m I_n - \lambda^{m-1} A_{m-1} - \cdots - \lambda A_1 - A_0$ with entrywise nonnegative coefficients $A_0, \ldots, A_{m-1}$ are called Perron polynomials. Following these authors, we will call P an m**–monic Perron-Frobenius polynomial** (PFP).

If there exists a monic right divisor C of P of degree m of the form

$$C(\lambda) = \lambda^m I_n - \sum_{j=0}^{m-1} \lambda^j C_j, \quad C_j \in \mathbb{C}^{n,n} \ (j \in \langle m-1 \rangle_0), \tag{4-46}$$

then there exists a unique $n \times n$-matrix polynomial B (see e.g. [Mar88]) with

$$B(\lambda) = \sum_{j=0}^{l-m} \lambda^j B_j \quad \text{and} \quad P(\lambda) = B(\lambda) C(\lambda). \tag{4-47}$$

We will call C in (4-46) and (4-47) a (right) **Perron-Frobenius factor** (PFF) **of degree** m of P if the coefficients C_j are entrywise nonnegative for all $j \in \langle m-1 \rangle_0$.

Since in this thesis we will only consider right PFF's, the word right will mostly be omitted.

If C is a PFF of P, then the spectral radius $\rho = \operatorname{spr} C$ of C is an eigenvalue of the polynomial C, since the companion matrix of C is a nonnegative matrix and has the same eigenvalues as C.

We are especially interested in factors for which the sets of eigenvalues are separated by a circle in the complex plane around zero. Following the notations in [Mar88], for $\rho > 0$ a right factor C of P is called a **spectral (right) factor with respect to** $\mathbb{T}_\rho = \{z \in \mathbb{C} \colon |z| = \rho\}$ if all eigenvalues of C lie inside the open disc $\mathbb{D}_\rho = \{z \in \mathbb{C} \colon |z| < \rho\}$ and the eigenvalues of B lie outside of $\overline{\mathbb{D}}_\rho$, i. e.

$$\sigma(C) = \sigma(P) \cap \mathbb{D}_\rho, \quad \text{and} \quad \sigma(B) = \sigma(P) \cap (\mathbb{C} \setminus \overline{\mathbb{D}}_\rho).$$

Recall that a matrix polynomial $Q(\lambda) = \sum_{j=0}^{l} \lambda^j Q_j$ is called **comonic** if $Q_0 = I_n$.

If C is a spectral right factor with respect to $\mathbb{T}_\rho$, 0 is not an eigenvalue of the left factor B and hence, the coefficient B_0 is nonsingular. Therefore, there exists a comonic matrix polynomial $\widetilde{B}$ with $\sigma(\widetilde{B}) = \sigma(B)$, such that

$$\widetilde{B}(\lambda) = I_n - \sum_{j=1}^{l-m} \lambda^j \widetilde{B}_j \quad \text{and} \quad P(\lambda) = \widetilde{B}(\lambda) B_0 C(\lambda).$$

Clearly, in this case, if λ_0 is an eigenvalue of P, it is either an eigenvalue of $\widetilde{B}$ or an eigenvalue of C. Hence, P has no eigenvalues on $\mathbb{T}_\rho$. This also implies that if P has a spectral right factor, then P is a regular matrix polynomial.

A more general version of the following theorem is proved in [FN05b], namely for coefficients which are in a normal cone of an ordered Banach algebra. The proof relies on an abstract factorization theorem in decomposing Banach algebras, see [GKS03, Theorem 2.5]. In section 4.2 we will give a proof in the matrix case using only matrix theoretical concepts.

Theorem 4.1. *(i) Let P be an m–monic PFP. Suppose that for some $\rho > 0$ the inequality*

$$\phi_A(\rho) < \rho^m$$

holds.

Then P has a spectral right PFF C with respect to the circle $\mathbb{T}_\rho$, i. e., $P(\lambda) = \widetilde{B}(\lambda)B_0C(\lambda)$ with

$$C(\lambda) = \lambda^m I_n - \sum_{j=0}^{m-1} \lambda^j C_j \quad \text{and} \quad \widetilde{B}(\lambda) = I_n - \sum_{j=1}^{l-m} \lambda^j \widetilde{B}_j,$$

where $C_i, \widetilde{B}_j$ are entrywise nonnegative for $i \in \langle m-1 \rangle_0$ and $j \in \langle l-m \rangle$ and $B_0 \in \mathbb{C}^{n,n}$ is a nonsingular M-matrix.

The sets of eigenvalues of C and $\widetilde{B}$ are

$$\sigma(C) = \sigma(P) \cap \mathbb{D}_\rho \text{ and } \sigma(\widetilde{B}) = \sigma(P) \cap (\mathbb{C} \setminus \overline{\mathbb{D}}_\rho).$$

(ii) Conversely, if the m–monic PFP has a spectral right PFF with respect to the circle $\mathbb{T}_\rho$ for some $\rho > 0$, then $\phi_A(\rho) < \rho^m$.

4.1 Factorizations of 1–monic matrix polynomials

In this section we study the existence of a spectral PFF for a 1–monic PFP. In Section 4.2 we will combine this result with the results of Chapter 3 to prove this theorem for any m–monic matrix polynomial with $1 \leqslant m < l$.

Consider a 1-monic PFP P of degree l,

$$P(\lambda) = \lambda I_n - A(\lambda) = \lambda I_n - \sum_{j=0}^{l} \lambda^j A_j. \tag{4-48}$$

We will give a sufficient condition for a PFP to have a spectral right PFF with respect to $\mathbb{T}_\rho$ for some $\rho > 0$, i. e.,

$$P(\lambda) = \tilde{B}(\lambda)B_0C(\lambda) = \left(I_n - \sum_{j=1}^{l-1} \lambda^j \tilde{B}_j \right) B_0(\lambda I_n - C_0), \tag{4-49}$$

with $C_0, \tilde{B}_j \geqslant 0$ for $j \in \langle l-1 \rangle$, $\sigma(C) = \sigma(P) \cap \mathbb{D}_\rho$, $\sigma(\tilde{B}) = \sigma(P) \cap (\mathbb{C} \setminus \overline{\mathbb{D}}_\rho)$ for some $\rho > 0$ and a nonsingular M-matrix B_0.

In the following we will study the nonnegative solutions of the matrix equation

$$P(X) = X - \sum_{j=0}^{l} A_j X^j = 0, \quad X \in \mathbb{C}^{n,n}. \tag{4-50}$$

By [Mar88, Lemma 22.9], the matrix C_0 is a solution of (4-50) if and only if the linear pencil $\lambda I_n - C_0$ is a right factor of P, i. e.

$$P(\lambda) = B(\lambda)(\lambda I_n - C_0), \tag{4-51}$$

where B is a matrix polynomial with $B(\lambda) = \sum_{j=0}^{l-1} \lambda^j B_j$.

We call a solution $C_0 \in \mathbb{R}^{n,n}$ of (4-50) a **right root** of P or of the matrix equation $P(X) = 0$.

C_0 is called a **spectral right root** of P with respect to $\mathbb{T}_\rho$ if $\lambda I_n - C_0$ is a spectral right factor of P with respect to $\mathbb{T}_\rho$.

Given a right root of P, the next lemma formulates an explicit formula for the coefficients of the left factor B of P, which we will need later.

Lemma 4.2. *Suppose that $C_0 \in \mathbb{C}^{n,n}$ is a right root of $P(\lambda) = \lambda I_n - \sum_{j=0}^{l} \lambda^j A_j$ and let*

$$P(\lambda) = \left(\sum_{j=0}^{l-1} \lambda^j B_j \right) (\lambda I_n - C_0) \tag{4-52}$$

be the according factorization of P. Then the coefficients B_k, $k \in \langle\, l-1 \,\rangle_0$ satisfy the equations

$$B_k = \delta_{0k} I_n - \sum_{j=0}^{l-k-1} A_{k+j+1} C_0^j \quad \text{for } k \in \langle\, l-1 \,\rangle_0 . \tag{4-53}$$

Proof. First write $P(\lambda) = \sum_{j=0}^{l} \lambda^j \tilde{A}_j$, where $\tilde{A}_j = \delta_{j1} I_n - A_j$ and consider the factorization (4-52).

By induction we will show that $B_{l-k} = \sum_{j=0}^{k-1} \tilde{A}_{l-k+j+1} C_0^j$ for $k \in \langle\, l \,\rangle$. Since this is equivalent to

$$\begin{aligned} B_k &= \sum_{j=0}^{l-k-1} \tilde{A}_{k+j+1} C_0^j = \sum_{j=0}^{l-k-1} \delta_{(k+j)0} I_n - \sum_{j=0}^{l-k-1} A_{k+j+1} C_0^j \\ &= \delta_{0k} I_n - \sum_{j=0}^{l-k-1} A_{k+j+1} C_0^j \quad \text{for } k \in \langle\, l-1 \,\rangle_0 , \end{aligned}$$

which proves the assertion.

By (4-52) it directly follows that $B_{l-1} = \tilde{A}_l$. Suppose that $B_{l-k} = \sum_{j=0}^{k-1} \tilde{A}_{l-k+j+1} C_0^j$ for some $k \in \langle\, l-1 \,\rangle$. By formally dividing $P(\lambda) = \sum_{j=0}^{l} \lambda^j \tilde{A}_j$ by $C(\lambda) = \lambda I_n - C_0$, performing a polynomial division, one can see that B_{l-k-1} is obtained by

$$\begin{aligned} B_{l-k-1} &= \tilde{A}_{l-k} - B_{l-k}(-C) = \tilde{A}_{l-k} + \left(\sum_{j=0}^{k-1} \tilde{A}_{l-k+j+1} C^j \right) C \\ &= \tilde{A}_{l-k} + \sum_{j=0}^{k-1} \tilde{A}_{l-k+j+1} C^{j+1} = \sum_{j=-1}^{k-1} \tilde{A}_{l-k+j+1} C^{j+1} \\ &= \sum_{j=0}^{k} \tilde{A}_{l-k+j} C^j . \end{aligned} \tag{4-54}$$

□

From identity (4-53) one can easily observe that the coefficients B_k of B have a certain structure if P has an entrywise nonnegative right root. The next corollary formulates this simple observation.

Corollary 4.3. *Suppose that $C \in \mathbb{C}^{n,n}$ is an entrywise nonnegative right root of the PFP (4-48). Then the coefficients B_k are nonpositive for $k \in \langle l-1 \rangle$ and B_0 is a Z-matrix.*

We will now try to find a solution of (4-50) via a fixpoint iteration.

We will say that a **sequence of $n \times n$ matrices $(X_k)_{k\in\mathbb{N}}$ converges** to an $n \times n$ matrix X if the n^2 sequences of components of X_k converge to the corresponding components of X. We will write $\lim_{k\to\infty} X_k = X$.

Recall the order relation induced by the closed algebra cone of entrywise nonnegative $n \times n$-matrices from Example 1.11.

Before we proof the next proposition, note that for entrywise nonnegative matrices $A = (a_{ij}), B = (b_{ij}) \in \mathbb{R}^{n,n}$ and $C = (c_{ij}), D = (d_{ij}) \in \mathbb{R}^{n,m}$ we have

$$AC \leqslant BD \quad \text{if} \quad A \leqslant B \text{ and } C \leqslant D. \tag{4-55}$$

Indeed, $(AC)_{ij} = \sum_{\nu=1}^{n} a_{i\nu} c_{\nu j} \leqslant \sum_{j=1}^{n} b_{i\nu} d_{\nu j} = (BD)_{ij}$ holds for all $i \in \langle n \rangle$ and $j \in \langle m \rangle$. In particular, this implies that for all $j \in \mathbb{N}$ we have that

$$A^j \leqslant B^j \quad \text{if} \quad A \leqslant B. \tag{4-56}$$

Proposition 4.4. *Let P be a 1-monic PFP of degree l. Then P has a nonnegative right root if and only if the fixpoint iteration*

$$X_{k+1} = A_l X_k^l + \cdots + A_1 X_k + A_0 \quad \textit{with } 0 \leqslant X_0 \leqslant A_0 \tag{4-57}$$

converges.

Furthermore, the limit is the entrywise smallest (or minimal) nonnegative root of P and does not depend on the choice of the initial matrix X_0.

Proof. Obviously $X_k \geqslant 0$ for all $k \geqslant 0$. By induction it follows that the sequence $(X_k)_{k\in\mathbb{N}}$ of iterates is entrywise nondecreasing. Indeed, since the coefficients A_j are nonnegative, clearly $X_0 \leqslant A_0 \leqslant X_1$. If we suppose that $X_k \geqslant X_{k-1}$ for some $k \geqslant 1$, then, due to (4-56),

$$X_{k+1} - X_k = A_l(X_k^l - X_{k-1}^l) + \cdots + A_1(X_k - X_{k-1}) + A_0.$$

Hence, if $(X_k)_{k\in\mathbb{N}}$ converges, the limit is entrywise nonnegative and the if part follows.

If we suppose that $X \in \mathbb{R}^{n,n}$ is any nonnegative right root of P, i.e., $X = A_l X^l + \cdots + A_1 X + A_0$, then again, by induction it follows that $X_k \leqslant X$ for all $k \in \mathbb{N}$, indeed $0 \leqslant X_0 \leqslant A_0 \leqslant X$ and if $X_k \leqslant X$ for some $k \geqslant 0$, then, due to (4-56),

$$X_{k+1} = \sum_{j=0}^{l} A_j X_k^j \leqslant \sum_{j=0}^{l} A_j X^j = X.$$

Hence, X is a bound for the entrywise monotonically increasing iteration sequence $(X_k)_{k\in\mathbb{N}}$. Hence, since the convergence is and boundedness is componentwise, $(X_k)_{k\in\mathbb{N}}$ converges.

Let X be any right root of P and $X_{\min}$ be the limit of (4-57) for any entrywise nonnegative initial matrix $X_0 \leqslant A_0$. From $X_k \leqslant X$ for all $k \in \mathbb{N}$ it follows that

$$X - X_{\min} = X - \lim_{k \to \infty} X_k \geqslant 0,$$

hence, $X_{\min}$ is the entrywise smallest nonnegative root of P. This also implies the last assertion on the independence of the initial matrix X_0.

□

Proposition 4.5. *Let P be a 1-monic PFP of degree l as in* (4-48). *Suppose that $A(\rho)v \leqslant \rho v$ for some $\rho > 0$ and some vector $v = (v_j)_{j=1}^n \in \mathbb{R}^n$, $v > 0$, i.e. $v_j > 0$ for all $j \in \langle n \rangle$. Then*

(i) P has a nonnegative right root.

(ii) the minimal nonnegative right root $C_{\min}$ of P satisfies

$$C_{\min} v \leqslant \rho v \quad \textit{and} \quad \operatorname{spr}(C_{\min}) \leqslant \rho.$$

Proof. (*i*) We show that the fixpoint iteration (4-57) converges. By induction one can see that $0 \leqslant X_k v \leqslant \rho v$ for all $k \in \mathbb{N}$. Indeed, since $v > 0$, $X_0 v \leqslant A_0 v \leqslant A(\rho)v \leqslant \rho v$ and if $X_k v \leqslant \rho v$ for some $k \geqslant 0$, then due to (4-55)

$$X_{k+1} v = \sum_{j=0}^{l} A_j X_k^j v \leqslant \sum_{j=0}^{l} A_j \rho^j v = A(\rho)v \leqslant \rho v.$$

Denote by $(X_k)_{ij}$ the element in the ith row and in the jth column of the matrix X_k. Then $\sum_{j=1}^n (X_k)_{ij} v_j \leqslant \rho v_i$ and hence, $(X_k)_{ij} \leqslant \rho \frac{v_i}{v_j}$ for all $i, j \in \langle n \rangle$. Hence, the iteration sequence $(X_k)_{k \in \mathbb{N}}$ is bounded by the matrix $\left(\rho \frac{v_i}{v_j}\right)$. In the proof of Proposition 4.4 we already saw that by definition and by the nonnegativity of the X_k, $(X_k)_{k \in \mathbb{N}}$ is entrywise nondecreasing. Hence, it converges and P has a nonnegative root.

(*ii*) $C_{\min} v \leqslant \rho v$ directly follows from $X_k v \leqslant \rho v$ (see the proof of (*i*)) for all $k \in \mathbb{N}$, since $C_{\min} = \lim_{k \to \infty} X_k$. [BP94, chapter 2, Theorem 1.11] then implies $\operatorname{spr}(C_{\min}) \leqslant \rho$.

□

The following example given in [FN05a] shows that that the assumption in Proposition 4.5 is not necessary for the existence of a nonnegative root of a PFP.

Example 4.6. Let $a, b > 0$ such that $ab > 1$ and define

$$P(\lambda) = \lambda I_2 - A(\lambda) = \left(I_2 - \lambda \begin{bmatrix} a & 0 \\ 0 & 0 \end{bmatrix} \right) \left(\lambda I_2 - \begin{bmatrix} 0 & 0 \\ 0 & b \end{bmatrix} \right).$$

Then for all for $\rho > 0$ and $v = \left[\begin{smallmatrix} x \\ y \end{smallmatrix}\right] > 0$ we have $P(\rho)v = \left[\begin{smallmatrix} (1-\rho a)\rho x \\ (\rho - b)y \end{smallmatrix}\right]$.

Now $1 - \rho a \geqslant 0$ implies that $a \leqslant \frac{1}{\rho}$, i.e., $\rho < b$, hence, $\rho - b < 0$. Conversely if $\rho - b \geqslant 0$, then $\rho a \geqslant ab > 1$, hence, $1 - \rho a < 0$. Therefore, there is no $\rho > 0$ and no $v > 0$ such that $A(\rho)v \leqslant \rho v$ but P has a nonnegative right root.

Proposition 4.7. *Let P be a 1-monic PFP of degree l as in (4-48). Suppose that there exists a $\rho > 0$ such that $\phi_A(\rho) = \operatorname{spr} A(\rho) < \rho$. Then P has a nonnegative right root.*

In this case, if $C_{\min}$ is the minimal entrywise nonnegative right root of P and

$$P(\lambda) = B(\lambda)(\lambda I_n - C_{\min}) = \left(\sum_{j=0}^{l-1} \lambda^j B_j\right)(\lambda I_n - C_{\min}) \tag{4-58}$$

is the according factorization of P, then the following statements hold.

(i) $\operatorname{spr}(C_{\min}) < \rho$,

(ii) B_0 *is a nonsingular M-matrix,*

(iii) the spectral radius of $\operatorname{rev}(BB_0^{-1})$ *is strictly less then* $\frac{1}{\rho}$,

(iv) $\sigma(C_{\min}) = \sigma(P) \cap \mathbb{D}_\rho$, $\quad \sigma(B) = \sigma(P) \cap (\mathbb{C} \setminus \overline{\mathbb{D}}_\rho)$,

i. e., $C_{\min}$ is a spectral right root of P.

Proof. Due to Proposition 1.7, by assumption ρ is not an eigenvalue of P and hence, $P(\rho)$ is invertible and $\operatorname{spr}\left(\frac{1}{\rho}A(\rho)\right) < 1$. Thus, we can write $P(\rho)^{-1}$ as a Neumann series (see e.g. [Kre89], [Wer00], [Heu06])

$$P(\rho)^{-1} = \frac{1}{\rho}\sum_{j=0}^{\infty}\left(\frac{1}{\rho}A(\rho)\right)^j \geqslant 0.$$

Take a vector $u \in \mathbb{R}^n$ with $u > 0$ and define $v = P(\rho)^{-1}u$, i.e., $P(\rho)v = u > 0$, i.e. $A(\rho)v < \rho v$. The fact that $P(\rho)^{-1}$ has full rank and is entrywise nonnegative implies $v > 0$. From Proposition 4.5 it follows that P has a nonnegative right root $C_{\min}$.

(*i*) From Proposition 4.5 it follows that $\operatorname{spr}(C_{\min}) \leqslant \rho$. Equality cannot hold here, because this would imply that there exists a vector $w \geqslant 0$, such that $C_{\min} w = \rho w$, since $C_{\min}$ is nonnegative. It follows that $P(\rho)u = B(\rho)(\rho I_n - C_{\min})w = 0$ in contradiction to $\operatorname{spr} A(\rho) < \rho$. Hence, $\operatorname{spr} C_{\min} < \rho$.

(*ii*) By Lemma 4.2 we have

$$B_0 = I_n - \sum_{j=0}^{l-1} A_{j+1} C_{\min}^j.$$

For the vector v from (*i*), by Proposition 4.5 (*ii*) and by (4-55) we have $C_{\min}^j v \leqslant \rho^j v$ for $j \in \langle l-1 \rangle$ and, therefore

$$\sum_{j=0}^{l-1} A_{j+1} C_{\min}^j v \leqslant \sum_{j=0}^{l-1} \rho^j A_{j+1} v = \frac{1}{\rho}\sum_{j=1}^{l} A_j \rho^j v \leqslant \frac{1}{\rho}\sum_{j=0}^{l} A_j \rho^j v < v,$$

where the last inequality is due to $P(\rho)v > 0$, see the proof of (*i*).

From [BP94, chapter 2, Theorem 1.11] it follows that $\operatorname{spr}(\sum_{j=0}^{l-1} A_{j+1} C_{\min}^j) < 1$, hence, due to Lemma 1.25, B_0 is invertible with a nonnegative inverse.

(*iii*) Due to $\operatorname{spr}(C_{\min}) < \rho$, (4-58) implies that $B(\rho) = P(\rho)(\rho I_n - C_{\min})^{-1}$ is invertible. Since $v \geqslant (\rho I_n - C_{\min})\frac{1}{\rho}v$, and, since $\rho I_n - C_{\min}$ has a nonnegative inverse, by (4-55) we have $(\rho I_n - C_{\min})^{-1}v \geqslant \frac{1}{\rho}v$ and hence

$$B(\rho)v = P(\rho)(\rho I_n - C_{\min})^{-1}v \geqslant \frac{1}{\rho}P(\rho)v > 0. \tag{4-59}$$

Set $\widetilde{B}_j = -B_j B_0^{-1}$, $(j \in \langle\, l-1\,\rangle)$ and define the matrix polynomial $\widetilde{B}$ via

$$\widetilde{B}(\lambda) = I_n - \sum_{j=1}^{l-1} \lambda^j \widetilde{B}_j.$$

Note that the coefficients $\widetilde{B}_j$, $j \in \langle\, l-1\,\rangle$ are nonnegative, since the B_j are nonpositive for all $j \in \langle\, l-1\,\rangle$ due to Corollary 4.3. In this notation $B(\lambda) = \widetilde{B}(\lambda)B_0$. Consider the monic matrix polynomial

$$\operatorname{rev}\widetilde{B}(\lambda) = \lambda^{l-1}I_n - \beta(\lambda) = \lambda^{l-1}I_n - (\lambda^{l-2}\widetilde{B}_1 + \cdots + \lambda\widetilde{B}_{l-2} + \widetilde{B}_{l-1}).$$

Since B_0 is a nonsingular M-matrix, (4-59) implies that $\widetilde{B}(\rho)v > 0$. Therefore, setting $\lambda = \frac{1}{\rho}$ we get

$$\begin{aligned}\operatorname{rev}\widetilde{B}(1/\rho)v &= \frac{1}{\rho^{l-1}}I_n - (\frac{1}{\rho^{l-2}}\widetilde{B}_1 + \cdots + \frac{1}{\rho}\widetilde{B}_{l-2} + \widetilde{B}_{l-1})v \\ &= \frac{1}{\rho^{l-1}}\left[I_n - (\rho\widetilde{B}_1 + \cdots + \rho^{l-2}\widetilde{B}_{l-2} + \rho^{l-1}\widetilde{B}_{l-1})\right]v \\ &= \frac{1}{\rho^{l-1}}\widetilde{B}(\rho)v > 0.\end{aligned}$$

This implies that $\operatorname{spr}\beta(1/\rho) < \frac{1}{\rho^{l-1}}$. Indeed, let $r = \operatorname{spr}\beta(1/\rho)$ and consider a vector $u \geqslant 0$, $u \neq 0$ such that $u^T\beta(1/\rho) = ru^T$. Then

$$0 < \left\langle \operatorname{rev}\widetilde{B}(1/\rho)v, u\right\rangle = \frac{1}{\rho^{l-1}}\langle v, u\rangle - \langle v, \beta(1/\rho)^T u\rangle = \left(\frac{1}{\rho^{l-1}} - r\right)\langle v, u\rangle,$$

hence, $r < \frac{1}{\rho^{l-1}}$.

Now apply Proposition 1.24 to the monic matrix polynomial $\operatorname{rev}\widetilde{B}$, $\operatorname{rev}\widetilde{B}(\lambda) = \lambda^{l-1}I_n - \beta(\lambda)$, i. e., from $\operatorname{spr}\beta(1/\rho) < \frac{1}{\rho^{l-1}}$ it follows that $\operatorname{spr}(\operatorname{rev}\widetilde{B}) < \frac{1}{\rho}$.

(*iv*) From (*i*) it follows that $\sigma(C_{\min}) \subset \mathbb{D}_\rho$. $\widetilde{B} = BB_0^{-1}$ is a comonic matrix polynomial, thus 0 is not an eigenvalue of $\widetilde{B}$. Since $0 \neq \lambda \in \sigma(\widetilde{B})$ if and only if $\frac{1}{\lambda} \in \sigma(\operatorname{rev}\widetilde{B})$, from (*iii*) it follows that $\sigma(\widetilde{B}) \cap \overline{\mathbb{D}}_\rho = \varnothing$. Furthermore, $\sigma(B) = \sigma(\widetilde{B})$ and from [Mar88, Section 22] the rest of the assertion follows.

□

Theorem 4.8. *Let P be a 1-monic PFP of degree l as in (4-48) and let $\rho > 0$. Then P has a spectral right PFF with respect to $\mathbb{T}_\rho$ if and only if $\operatorname{spr}A(\rho) < \rho$.*

Moreover if C_0 is a spectral right root of P with respect to $\mathbb{T}_\rho$, then it is the entrywise smallest nonnegative root $C_{\min}$ of P.

Proof. The if part follows from Proposition 4.7.

Suppose that P has a spectral right PFF, i. e., there is a matrix polynomial $\widetilde{B}$ and matrices C_0 and B_0 such that

$$P(\lambda) = \widetilde{B}(\lambda) B_0 C(\lambda) = \Big(I_n - \sum_{j=1}^{l-1} \lambda^j \widetilde{B}_j\Big) B_0(\lambda I_n - C_0), \tag{4-60}$$

where $C_0, \widetilde{B}_1, \ldots, \widetilde{B}_{l-1}$ are nonnegative matrices and B_0 is a nonsingular M–matrix and

$$\sigma(C) = \sigma(P) \cap \mathbb{D}_\rho, \quad \sigma(\widetilde{B}) = \sigma(P) \cap (\mathbb{C} \setminus \overline{\mathbb{D}}_\rho).$$

So, in particular, $\operatorname{spr}(C) < \rho$ and $\operatorname{spr}(\operatorname{rev} \widetilde{B}) < \frac{1}{\rho}$. Together with Proposition 1.24, this implies that if

$$\operatorname{rev} \widetilde{B}(\lambda) = \lambda^{l-1} I_n - \beta(\lambda) = \lambda^{l-1} I_n - (\lambda^{l-2} \widetilde{B}_1 + \ldots \lambda \widetilde{B}_{l-2} + \widetilde{B}_{l-1}),$$

then $\operatorname{spr} \beta(\frac{1}{\rho}) < \frac{1}{\rho^{l-1}}$. By multiplication with ρ^{l-1} we obtain

$$\operatorname{spr}(\widetilde{B}(\rho)) = \operatorname{spr}(\rho \widetilde{B}_1 + \cdots + \rho^{l-1} \widetilde{B}_{l-1}) < 1.$$

Hence, $\widetilde{B}(\rho)$ has a nonnegative inverse. Due to $\operatorname{spr}(C) = \operatorname{spr}(C_0) < \rho$, also $(\rho I_n - C_0)$ is an invertible M-matrix. We have

$$P(\rho) = \Big(I_n - \sum_{j=1}^{l-1} \rho^j \widetilde{B}_j\Big) B_0(\rho I_n - C_0),$$

where B_0 is an invertible M-matrix. Therefore, $P(\rho) = \rho I_n - A(\rho)$ is a nonsingular M-matrix and [BP94, p. 137, (N_{38})] implies that $\operatorname{spr} A(\rho) < \rho$.

The last assertion is verified with Proposition 4.7, since it implies that $\sigma(C_{\min}) = \sigma(P) \cap \sigma(\mathbb{D}_\rho) = \sigma(C)$ and from [Mar88, Lemma 22.8] it follows that $C_0 = C_{\min}$. □

4.2 Factorizations of m–monic matrix polynomials

The main aim of this section is to prove Theorem 4.1, which has already been done in Section 4.1 for the case of a 1-monic PFP. We will combine the results about 1–monic PFFs in Section 4.1 and results from Chapter 3 to prove the theorem.

Lemma 4.9. *Suppose that the m–monic PFP P has a right PFF, i. e.,*

$$P(\lambda) = \left(\sum_{j=0}^{l-m} \lambda^j B_j\right)\left(\lambda^m I_n - \sum_{j=0}^{m-1} \lambda^j C_j\right) \quad \textit{with} \quad C_j \geqslant 0 \quad (j \in \langle m-1 \rangle_0).$$

Then B_0 is a Z-matrix.

Proof. For the case $m = 1$, see Section 4.1.

By Corollary 3.6, the canonical reduction $\mathcal{P}$ of P has a nonnegative right root $\mathcal{C}$, i. e. $\mathcal{P}(\lambda) = \mathcal{B}(\lambda)(\lambda I_{mn} - \mathcal{C})$. Let $\mathcal{B}(\lambda) = \sum_{j=0}^{l-m} \lambda^j \mathcal{B}_j$, then Corollary 4.3 implies that the coefficient $\mathcal{B}_0$ is a Z-matrix. From the form of $\mathcal{B}_0$, see (3-40) in the proof of Theorem 3.4 it then follows that also B_0 is a Z-matrix. □

In the next Proposition we will use the notion of irreducible matrices. Following [HJ85, Sec. 6.2] we define that a matrix $S \in \mathbb{C}^{n,n}$ is **irreducible** if there exists no permutation matrix $P \in \mathbb{C}^{n,n}$ such that there is some integer r with $1 \leqslant r \leqslant n-1$ such that

$$P^T SP = \begin{bmatrix} B & C \\ 0 & D \end{bmatrix},$$

where $B \in \mathbb{C}^{r,r}$, $C \in \mathbb{C}^{r,n-r}$, $D \in \mathbb{C}^{n-r,n-r}$.

From the Perron-Frobenius theory it follows that if S is an irreducible matrix, then $\operatorname{spr}(S)$ is a simple eigenvalue of S, i. e. the algebraic multiplicity of $\operatorname{spr}(S)$ as an eigenvalue of S is one (and so is the dimension of the corresponding eigenspace), and there is a strictly positive eigenvector of S corresponding to $\operatorname{spr}(S)$ (see e.g. [HJ91], [Min88] and for a generalization to infinite dimensions see [Sch71]).

Recall the function

$$\Lambda : \mathbb{C} \to L(\mathbb{C}^n, \mathbb{C}^{mn}), \quad \Lambda(\lambda)u = \begin{bmatrix} \lambda^{m-1} \\ \vdots \\ \lambda \\ 1 \end{bmatrix} \otimes u, \quad \text{for } u \in \mathbb{C}^n.$$

from (3-45), where $l - k + 1 = m$.

Proposition 4.10. *Let $m \geqslant 2$ and let P be an m–monic PFP with $P(\lambda) = \lambda^m I_n - A(\lambda) = \lambda^m I_n - \sum_{j=0}^{l} \lambda^j A_j$. Consider the canonical reduction $\mathcal{P}$ of P with $\mathcal{P}(\lambda) = \lambda I_{mn} - \mathcal{A}(\lambda)$ according to Corollary 3.3. Then for $\rho > 0$ we have that*

$$\phi_A(\rho) < \rho^m \quad \text{if and only if} \quad \phi_{\mathcal{A}}(\rho) < \rho, \tag{4-61}$$

$$\phi_A(\rho) > \rho^m \quad \text{if and only if} \quad \phi_{\mathcal{A}}(\rho) > \rho \tag{4-62}$$

and

$$\phi_A(\rho) = \rho^m \quad \text{if and only if} \quad \phi_{\mathcal{A}}(\rho) = \rho. \tag{4-63}$$

Proof. We first prove both statements for the case that for all positive τ the matrix $A(\tau)$ is irreducible.

Let $\phi_A(\rho) = \rho^m$. Since $A(\rho)$ is irreducible, there exists a positive eigenvector $x > 0$ of $A(\rho)$ to the eigenvalue $\rho^m = \operatorname{spr} A(\rho)$. Hence, $P(\rho)x = 0$. By Proposition 3.9 the positive vector $\Lambda(\rho)x$ is an eigenvector for $\mathcal{P}$ to the eigenvalue ρ. Hence, $\mathcal{A}(\rho)\Lambda(\rho)x = \rho\Lambda(\rho)x$. Since $\mathcal{A}(\rho)$ is entrywise nonnegative, we conclude that $\rho = \operatorname{spr} \mathcal{A}(\rho) = \phi_{\mathcal{A}}(\rho)$ (see e.g. [HJ85, Corollary 8.1.30]).

Conversely, suppose that $\phi_{\mathcal{A}}(\rho) = \rho$. Since $\mathcal{A}(\rho) \geqslant 0$, there exists a nonzero vector $v \geqslant 0$ such that $\mathcal{A}(\rho)v = \rho v$, which is equivalent to $\mathcal{P}(\rho)v = 0$. By Proposition 3.9 v is of the form $v = \Lambda(\rho)u$, where $u \geqslant 0$, $u \neq 0$ is an eigenvector of P to the same eigenvalue ρ, hence, $A(\rho)u = \rho^m u$. Since the matrix $A(\rho)$ is irreducible, it has only one nonnegative - hence, a strictly positive - eigenvector (except for scalar multiples) and this corresponds to the eigenvalue $\operatorname{spr} A(\rho)$. Therefore, $\phi_A(\rho) = \operatorname{spr} A(\rho) = \rho^m$.

Now suppose that $\operatorname{spr} A(\rho) < \rho^m$ and let $v > 0$ such that $A(\rho)v = \operatorname{spr} A(\rho)v < \rho^m v$. Set $\hat{v} = \Lambda(\rho)v > 0$ and consider

$$\mathcal{A}(\rho)\hat{v} = \begin{bmatrix} A(\rho)v \\ \rho^{m-1}v \\ \vdots \\ \rho v \end{bmatrix} \leqslant \begin{bmatrix} \rho^m v \\ \rho^{m-1}v \\ \vdots \\ \rho v \end{bmatrix} = \rho\hat{v},$$

which implies that $\operatorname{spr}\mathcal{A}(\rho) \leqslant \rho$, see e.g. [HJ85, Corollary 8.1.29]. Due to (4-63), equality cannot hold, so $\phi_{\mathcal{A}}(\rho) < \rho$. Analogously one can prove that $\phi_{\mathcal{A}}(\rho) > \rho$ when $\phi_A(\rho) > \rho^m$. Then it follows that $\phi_A(\rho) \geqslant \rho^m$ and $\phi_A(\rho) \leqslant \rho^m$ if $\phi_{\mathcal{A}}(\rho) \geqslant \rho$ and $\phi_{\mathcal{A}}(\rho) \leqslant \rho$, respectively. Analogously, with (4-63) the remainder follows and (4-61) and (4-62) are proved.

Now suppose that P is any m–monic PFP such that for $\tau > 0$, $A(\tau)$ is not necessarily irreducible. Define $A_\epsilon(\lambda) = A(\lambda) + \epsilon \mathbf{1}_{n,n}$, where $\mathbf{1}_{n,n}$ denotes the $n \times n$ matrix each entry of which is 1.

Let $\phi_A(\rho) < \rho^m$. Since the eigenvalues of A_ϵ depend continuously on ϵ and $A_\epsilon(\lambda) \to A(\lambda)$ entrywise as $\epsilon \to 0$, there is an $\epsilon > 0$ such that $\phi_{A_\epsilon}(\rho) < \rho^m$.

Consider the canonical reduction $\mathcal{P}_\epsilon(\lambda) = \lambda I_n - \mathcal{A}_\epsilon(\lambda)$ of $\lambda^m I_n - A_\epsilon(\lambda)$ according to Corollary 3.3. Since A_ϵ is irreducible, from the first part of this proof it follows that $\phi_{\mathcal{A}_\epsilon}(\rho) < \rho$. Since the coefficient $\mathcal{A}_{0,\epsilon}$ of $\mathcal{A}_\epsilon$ satisfies

$$\mathcal{A}_{0,\epsilon} = \begin{bmatrix} A_{m-1} & \cdots & \cdots & A_0 + \epsilon \mathbf{1}_{n,n} \\ I_n & & & 0 \\ & \ddots & & \vdots \\ & & I_n & 0 \end{bmatrix}, \tag{4-64}$$

we have $\phi_{\mathcal{A}}(\rho) = \lim_{\epsilon \to 0} \phi_{\mathcal{A}_\epsilon}(\rho) \leqslant \rho$. Due to $\phi_A(\rho) < \rho^m$, ρ is not an eigenvalue of P and, therefore it is not an eigenvalue of $\mathcal{P}$. Therefore, $\phi_{\mathcal{A}}(\rho) \neq \rho$, and hence, $\phi_{\mathcal{A}}(\rho) < \rho$.

Now suppose that $\phi_{\mathcal{A}}(\rho) < \rho$ and let $\epsilon > 0$ be such that $\phi_{\mathcal{A}_\epsilon}(\rho) < \rho$, where $\mathcal{A}_\epsilon$ coincides with $\mathcal{A}$ except for the coefficient $\mathcal{A}_0$, which is replaced by $\mathcal{A}_{0,\epsilon}$ from (4-64). $\mathcal{P}_\epsilon(\lambda) = \lambda I_{mn} - \mathcal{A}_\epsilon(\lambda)$ is the canonical reduction of $\lambda^m I_n - A_\epsilon(\lambda)$, where $A_\epsilon(\tau)$ is irreducible for all $\tau > 0$. So by the first part of this proof, $\phi_{A_\epsilon}(\rho) < \rho^m$. Again, $\lim_{\epsilon - 0} \phi_{A_\epsilon}(\rho) = \phi_A(\rho)$, so $\phi_A(\rho) \leqslant \rho^m$ and due to $\phi_{\mathcal{A}}(\rho) < \rho^m$, ρ is not an eigenvalue of $\mathcal{P}$, hence, it is not an eigenvalue of P, therefore $\phi_A(\rho) < \rho^m$, since $A(\rho)$ is entrywise nonnegative. The proof of the equivalence with " $>$ " works completely analogous. The last statement then follows immediately. ☐

We now summarize the previous observations in the theorem from the beginning of this section, Theorem 4.1, and will give a prove.

Theorem 4.1. *(i) Let P be an m–monic PFP with*

$$P(\lambda) = \lambda^m I_n - A(\lambda) = \lambda^m I_n - \sum_{j=0}^{l} \lambda^j A_j.$$

Suppose that for some $\rho > 0$ the inequality

$$\phi_A(\rho) < \rho^m$$

holds.

Then P has a spectral right Perron-Frobenius factor C with respect to the circle $\mathbb{T}_\rho$, i. e., $P(\lambda) = \widetilde{B}(\lambda) B_0 C(\lambda)$ with

$$C(\lambda) = \lambda^m I_n - \sum_{j=0}^{m-1} \lambda^j C_j \quad and \quad \widetilde{B}(\lambda) = I_n - \sum_{j=1}^{l-m} \lambda^j \widetilde{B}_j,$$

where $C_i, \widetilde{B}_j$ are entrywise nonnegative for $i \in \langle m-1 \rangle_0$ and $j \in \langle l-m \rangle$ and $B_0 \in \mathbb{C}^{n,n}$ is a nonsingular M-matrix.

Furthermore,

$$\sigma(C) = \sigma(P) \cap \mathbb{D}_\rho \text{ and } \sigma(\widetilde{B}) = \sigma(P) \cap (\mathbb{C} \setminus \overline{\mathbb{D}}_\rho).$$

(*ii*) *Conversely, if the m-monic PFP has a spectral right Perron-Frobenius factor with respect to the circle $\mathbb{T}_\rho$ for some $\rho > 0$, then $\phi_A(\rho) < \rho^m$.*

Proof. (*i*) For $m = 1$, see Section 4.1. By Proposition 4.10 $\phi_{\mathcal{A}}(\rho) < \rho$. From Theorem 4.8 it then follows that $\mathcal{P}$ has a nonnegative spectral right root $\mathcal{C}$ with respect to $\mathbb{T}_\rho$, i. e.,

$$\mathcal{P}(\lambda) = \widetilde{\mathcal{B}}(\lambda)\mathcal{B}_0(\lambda I_{mn} - \mathcal{C}) = \left(I_{mn} - \sum_{j=1}^{l-m} \lambda^j \widetilde{\mathcal{B}}_j\right) \mathcal{B}_0(\lambda I_{mn} - \mathcal{C}),$$

with $\sigma(\mathcal{C}) = \sigma(\mathcal{P}) \cap \mathbb{D}_\rho$ and $\sigma(\widetilde{\mathcal{B}}) = \sigma(\mathcal{P}) \cap (\mathbb{C} \setminus \overline{\mathbb{D}}_\rho)$ and $\mathcal{B}_0$ is a nonsingular M-matrix. Corollary 3.6 and its proof imply that P has a right divisor C such that

$$P(\lambda) = B(\lambda)C(\lambda) = \left(\sum_{j=0}^{l-m} \lambda^j B_j\right)\left(\lambda^m I_n - \sum_{j=0}^{m-1} \lambda^j C_j\right),$$

where due to (3-40) for B_0 we have

$$\mathcal{B}_0 = \begin{bmatrix} B_0 & * & * & * \\ & I_n & & \\ & & \ddots & \\ & & & I_n \end{bmatrix}.$$

B_0 is nonsingular, since $\mathcal{B}_0$ is nonsingular, and so by setting $\widetilde{B}_j = -B_j B_0^{-1}$ for $j \in \langle\, l-m \,\rangle$ we obtain

$$P(\lambda) = \left(I_n - \sum_{j=1}^{l-m} \lambda^j \widetilde{B}_j\right) B_0 \left(\lambda^m I_n - \sum_{j=0}^{m-1} \lambda^j C_j\right),$$

where B_0 is a nonsingular M-matrix due to the fact that $\mathcal{B}_0$ is a nonsingular M-matrix and $\mathcal{B}_0^{-1} = \begin{bmatrix} B_0^{-1} & * & * & * \\ & I_n & & \\ & & \ddots & \\ & & & I_n \end{bmatrix}$. Therefore, the coefficients $\widetilde{B}_j$ ($j \in \langle\, l-m \,\rangle$) are nonnegative.

Since $\mathcal{C}$ is the companion matrix of C, we have a splitting of the eigenvalues of P according to that of $\mathcal{P}$, i. e. $\sigma(C) = \sigma(P) \cap \mathbb{D}_\rho$ and $\sigma(\widetilde{B}) = \sigma(P) \cap (\mathbb{C} \setminus \overline{\mathbb{D}}_\rho)$

(*ii*) If conversely P has a spectral right PFF with respect to $\mathbb{T}_\rho$, its canonical reduction $\mathcal{P}$ has a spectral right root with respect to $\mathbb{T}_\rho$ and by Theorem 4.8 we have $\phi_{\mathcal{A}}(\rho) < \rho$, thus $\phi_A(\rho) < \rho^m$ by Proposition 4.10.

□

A quick glance on the connection between the right root $\mathcal{C}$ of $\mathcal{P}$ and the corresponding right PFF of P, see Corollary 3.6(*ii*), verifies the following proposition.

Proposition 4.11. *If there exists a PFF for an m–monic PFP then it also has a* **minimal right Perron-Frobenius factor**

$$\tilde{C}(\lambda) = \lambda^m I_n - \sum_{j=0}^{m-1} \lambda^j \tilde{C}_j$$

in the sense that if $C(\lambda) = \lambda^m I_n - \sum_{j=0}^{m-1} \lambda^j C_j$ is another PFF, then $\tilde{C}_j \leqslant C_j$ for all $j \in \langle m-1 \rangle$.

4.3 Nonnegative irreducible matrix polynomials

In this section we investigate spectral properties of m–monic PFP's

$$P(\lambda) = \lambda^m I_n - A(\lambda) = \lambda^m I_n - \sum_{j=0}^{l} \lambda^j A_j, \tag{4-65}$$

$m \geqslant 1$, where the polynomial A satisfies some irreducibility condition. Let $\rho, \tau \in \mathbb{R}$, $0 < \rho \leqslant \tau$. Then $\sum_{j=0}^{l} \tau^{j-l} A_j \leqslant \sum_{j=0}^{l} \rho^{j-l} A_j$, hence

$$0 \leqslant \frac{\rho^l}{\tau^l} A(\tau) \leqslant A(\rho) \leqslant A(\tau) \leqslant \frac{\tau^l}{\rho^l} A(\rho).$$

Therefore, the matrix $A(\tau)$ is irreducible for one positive τ if and only if $A(\tau)$ is irreducible for all positive τ. We define the matrix polynomial A to be **irreducible** if $A(\tau)$ is irreducible for all $\tau > 0$.

Due to the simplicity of $r_\tau = \operatorname{spr} A(\tau)$ as an eigenvalue of $A(\tau)$ we have

$$\mathsf{N}\big(r_\tau I_n - A(\tau)\big) \dotplus \mathsf{R}\big(r_\tau I_n - A(\tau)\big) = \mathbb{C}^n,$$

since if we suppose that there exists an $0 \neq y \in \mathsf{N}\big(r_\tau I_n - A(\tau)\big) \cap \mathsf{R}\big(r_\tau I_n - A(\tau)\big)$, then there exists a nonzero $x \in \mathbb{C}^n$ such that $(r_\tau I_n - A(\tau))x = y$, i. e., $x \in \mathsf{N}(r_\tau I_n - A(\tau))^2$, which cannot be the case, since r_τ is an algebraically simple eigenvalue.

Take $u_\tau, v_\tau > 0$ such that $A(\tau)u_\tau = r_\tau u_\tau$, $A(\tau)^T v_\tau = r_\tau v_\tau$ and $u_\tau^T v_\tau = 1$. Then $E(\tau) = u_\tau v_\tau^T$ is the spectral projection of $A(\tau)$ mapping from $\mathbb{C}^n$ onto $\mathsf{N}\big(r_\tau I_n - A(\tau)\big)$ along $\mathsf{R}\big(r_\tau I_n - A(\tau)\big) = \mathsf{N}(r_\tau I_n - A^T(\tau))^\perp$. Obviously, $E(\tau)$ is strictly positive, and has rank 1.

The analytic perturbation theory of eigenvalues (see [Bau85, pp. 93, 113, 144], [Kat76, II - 1, 2]) shows that the maps

$$\begin{aligned} \phi_A : (0,\infty) \to [0,\infty) \quad &\text{with} \quad \tau \longmapsto \operatorname{spr} A(\tau) \quad \text{and} \\ E : (0,\infty) \to \mathbb{R}^{n,n} \quad &\text{with} \quad \tau \longmapsto E(\tau) \end{aligned}$$

are real analytic. Furthermore, by Proposition 1.14, $\phi_A(\rho) = \max_{|\lambda|=\rho} \operatorname{spr} A(\lambda)$, thus ϕ_A is geometrically convex.

Proposition 4.12. *Let P be an m–monic PFP of degree l with irreducible A. Then either $\phi_A(\rho) = \rho^m$ for all nonnegative ρ or there are at most two positive ρ with $\phi_A(\rho) = \rho^m$.*

Proof. Suppose there are $0 < \rho_1 < \rho < \rho_2$ such that $\phi_A(\rho_1) = \rho_1^m$, $\phi_A(\rho) = \rho^m$, $\phi_A(\rho_2) = \rho_2^m$. Then by Proposition 1.4, $\phi_A(\tau) = \tau^m$ for all $\tau \in [\rho_1, \rho_2]$ and hence, since the functions ϕ_A and $\tau \mapsto \tau^m$ are analytic on $(0, \infty)$, we have $\phi_A(\tau) = \tau^m$ for all $\tau \in (0, \infty)$. □

Consider $u, v \in \mathbb{R}^n \setminus \{0\},\ u, v \geqslant 0,\ \tau > 0$. Differentiating the identity

$$\langle \phi_A(\tau)E(\tau)u, E^T(\tau)v\rangle = \langle A(\tau)E(\tau)u, E^T(\tau)v\rangle$$

with respect to τ yields

$$\begin{aligned}&\phi_A'(\tau)\left\langle E(\tau)u, E(\tau)^T v\right\rangle + \phi_A(\tau)\left\langle E'(\tau)u, E^T(\tau)v\right\rangle + \phi_A(\tau)\left\langle E(\tau)u, E'(\tau)^T v\right\rangle \\ &= \left\langle A'(\tau)E(\tau)u, E(\tau)^T v\right\rangle + \left\langle A(\tau)E'(\tau)u, E^T(\tau)v\right\rangle + \left\langle A(\tau)E(\tau)u, E'(\tau)^T v\right\rangle.\end{aligned} \tag{4-66}$$

Notice that for $u > 0$ and $\tau > 0$ the vector $E(\tau)u$ is a strictly positive eigenvector of $A(\tau)$ corresponding to its spectral radius $\operatorname{spr} A(\tau)$, and $E^T(\tau)u$ is a strictly positive eigenvector of $A^T(\tau)$ to $\operatorname{spr} A(\tau)$. Hence, (4-66) reduces to

$$\phi_A'(\tau)\left\langle E(\tau)u, E(\tau)^T v\right\rangle = \left\langle A'(\tau)E(\tau)u, E(\tau)^T v\right\rangle$$

and we obtain a representation of ϕ_A'.

$$\phi_A'(\tau) = \frac{\left\langle A'(\tau)E(\tau)u, E^T(\tau)v\right\rangle}{\langle E(\tau)u, E^T(\tau)v\rangle} \quad \text{for} \quad \tau > 0, \quad u, v \in \mathbb{R}^n \setminus \{0\},\ u, v \geqslant 0. \tag{4-67}$$

Proposition 4.13. *Let P be an m–monic PFP of degree l with irreducible A. Then the following assertions hold.*

(*i*) *Let $P(\rho)u = 0$ for some $\rho > 0$ and a nonzero vector $u \geqslant 0$.*
Then $\rho^m = \operatorname{spr} A(\rho), u > 0$ and $\mathsf{N}(P(\rho)) = \operatorname{span}\{u\}$.

(*ii*) *Let $P(\omega\rho)x = 0$ for some ρ with $\operatorname{spr} A(\rho) = \rho^m > 0$, $|\omega| = 1$ and $x \neq 0$.*
Then $P(\rho)|x| = 0$ and $|x| > 0$.

(*iii*) $\dim \mathsf{N}(P(\omega\rho)) \leqslant 1$ *for $\rho^m = \operatorname{spr} A(\rho) > 0$ and $|\omega| = 1$.*

Proof. (*i*) $P(\rho)u = 0$ means that ρ^m is an eigenvalue of $A(\rho)$ and u is a corresponding nonnegative eigenvector. Since $A(\rho)$ is irreducible, it has exactly one nonnegative eigenvector except for scalar multiples (see [Min88, Theorem 4.4]). By the Perron Frobenius Theorem this eigenvector corresponds to the eigenvalue $\operatorname{spr} A(\rho)$, i. e., $\rho^m = \operatorname{spr} A(\rho)$ and we have $u > 0$. Furthermore, $\operatorname{spr} A(\rho)$ is a simple eigenvalue of $A(\rho)$ which implies the last assertion.

(*ii*) From $P(\omega\rho)x = 0$ it follows that $\rho^m|x| = |A(\omega\rho)x| \leqslant A(\rho)|x|$. $A(\rho)$ is irreducible and $\rho^m = \operatorname{spr} A(\rho)$. Hence, by a completely analogous argumentation as in [Min88, p.12] we conclude that $A(\rho)|x| = \rho^m|x|$.

(*iii*) Let $x = \left[\begin{smallmatrix} x_1 \\ x_2 \end{smallmatrix}\right]$ and $y = \left[\begin{smallmatrix} y_1 \\ y_2 \end{smallmatrix}\right]$ be nonzero vectors in $\mathsf{N}(P(\omega\rho))$. From (*ii*) it follows that $x_1, x_2, y_1, y_2 \neq 0$. Then $y_1 x - x_1 y$ is a vector in $\mathsf{N}(P(\omega\rho))$ with 0 as its first component. By (*ii*) this implies that $y_1 x - x_1 y = 0$, hence, x and y are linearly dependent.

□

We will characterize the number of eigenvalues of P on the circles $\mathbb{T}_\rho$ such that $\rho^m = \operatorname{spr} A(\rho) > 0$ in a similar way as it is known for the number of peripheral eigenvalues of an irreducible matrix. We use some graph theoretical concepts used in [GHT96, Sec. 4] to study the spectral properties of certain Markov chains.

First, we briefly recall the concept of the directed graph associated with an entrywise nonnegative matrix, see e.g. [BP94], [HJ85]. Define the **associated directed graph** G_A of an $n \times n$ matrix A as the graph with n vertices, $V = \{1, \ldots, n\}$, such that for $i, j \in \langle n \rangle$ there is a directed edge from i to j if and only if $A_{ij} \neq 0$. Denote by $E = \{(i,j) \colon A_{ij} \neq 0\}$ the set of edges of G_A. We write $G_A = (V, E)$. We say that a sequence $(j_1, j_2), (j_2, j_3), \ldots, (j_{k-1}, j_k), (j_k, j_{k+1})$ of edges in E is a **path of** length k in G_A connecting j_1 and j_{k+1}. For simplicity we also write $j_1 \to j_2 \to \cdots \to j_k \to j_{k+1}$ or $(j_r)_{r=1}^{k+1}$, as appropriate. It is well known that an entrywise nonnegative matrix A is irreducible if and only if its associated directed graph G_A is **strongly connected**, i. e. that for each pair $i, j \in V$ there exists a path in G_A leading from i to j, see e.g. [BP94], [HJ85].

We now introduce an infinite graph associated with an m-monic matrix polynomial $P(\lambda) = \lambda^n I_n - \sum_{\nu=0}^{l} \lambda^\nu A_\nu$. Consider the infinite graph $G_m(A_0, A_1, \ldots, A_l) = G = (V, E)$ with the set of vertices and edges

$$
\begin{aligned}
V &= \{(j,p) \mid 1 \leqslant j \leqslant n, p \in \mathbb{Z}\} \quad \text{and} \\
E &= \{[(j,p),(k,q)] \mid A_{m+q-p}(j,k) > 0\}, \text{ respectively,}
\end{aligned}
$$

where $A_\nu(j,k)$ denotes the entry in the j-th row and the k-th column of the $n \times n$ matrix A_ν.

According to [GHT96] we call j the **phase** and p the **level** of $(j,p) \in V$.

Note that any level of the infinite graph $G_0(A_0)$ coincides with G_{A_0}.

Analogously to a finite graph, a sequence of subsequent edges is called a path in G. For a path

$$
(j_1, p_1) \to (j_2, p_2) \to \cdots \to (j_{s+1}, p_{s+1})
$$

in G the number

$$
p_{s+1} - p_1 = \sum_{r=1}^{s} (\nu_r - m), \quad \text{where } A_{\nu_r}(j_r, j_{r+1}) > 0
$$

is called the **level displacement** of the path.

Example 4.14. Consider $P(\lambda) = \lambda^3 I_8 - A(\lambda) = \lambda^3 I_8 - \sum_{j=0}^{4} \lambda^j A_j$, where for $j \in \langle 4 \rangle_0$, the coefficients $A_j \in \mathbb{R}^{8,8}$ are such that their only nonzero entries are

$$
\begin{aligned}
&A_1(7,8), \\
&A_2(4,6), \\
&A_3(3,1), \quad A_3(2,4), \quad A_3(8,5), \quad A_3(6,7). \\
&A_4(1,2), \quad A_4(5,3).
\end{aligned}
$$

Then the graph $G_3(A_0, A_1, A_2, A_3, A_4)$ of A is shown in the following image:

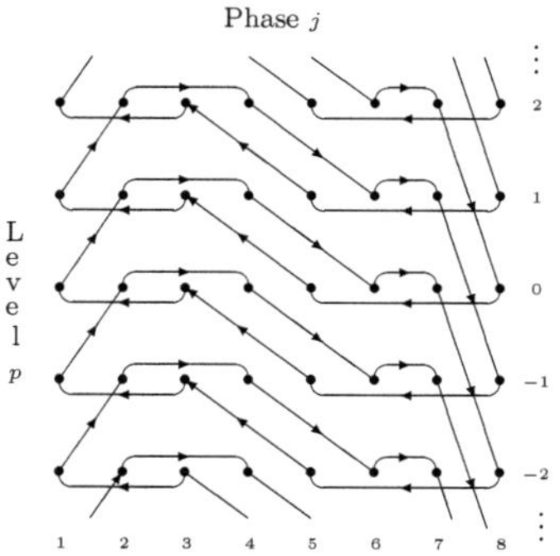

Figure 2: $G_3(A_0, A_1, A_2, A_3, A_4)$

The path $(5, -1) \to (3, 0) \to (1, 0) \to (2, 1)$, for instance, has the level displacement 2.

Remark 4.15. *Note that for the graph $G_m(A_0, A_1, \dots, A_l) = (V, E)$ we have that $[(j, p), (k, q)] \in E$ is equivalent to $[(j, p+u), (k, q+u)] \in E$ for all $u \in \mathbb{Z}$. This implies that if $u \in \mathbb{Z}$ and $((j_r, p_r))_{r=1}^{s}$ is a path in $G_m(A_0, A_1, \dots, A_l)$, then $((j_r, p_r + u))_{r=1}^{s}$ is a path in $G_m(A_0, A_1, \dots, A_l)$ and both paths have the same level displacement.*

We study some properties of the infinite graph G.

Lemma 4.16. *Let P be an m–monic PFP of degree l with irreducible A. For $j, k \in \langle\, n\,\rangle$ the following assertions are equivalent:*

(i) There exists a path from j to k in the associated directed graph $G_{A(1)}$ of the matrix $A(1)$.

(ii) For all $p \in \mathbb{Z}$ there exists a $q \in \mathbb{Z}$ such that there is a path from (j, p) to (k, q) in $G_m(A_0, A_1, \dots, A_l)$.

(iii) For all $q \in \mathbb{Z}$ there exists a $p \in \mathbb{Z}$ such that there is a path from (j, p) to (k, q) in $G_m(A_0, A_1, \dots, A_l)$.

All these paths in (i) to (iii) can be chosen to have the same number of vertices.

Proof. Suppose that (i) holds and let $j = j_0 \to j_1 \to \cdots \to j_{s+1} = k$ be a path of length s in $G_{A(1)}$. There exist $s \in \mathbb{N}$ and $j_1, \dots, j_{s+1} \in \langle\, n\,\rangle$ such that $j_1 = j$, $j_{s+1} = k$ and $A(1)(j_r, j_{r+1}) > 0$ for all $r \in \langle\, s\,\rangle$. Due to the nonnegativity of the coefficients $A_0, \dots, A_l$, for each $r \in \langle\, s\,\rangle$ there exists some $\nu_r \in \langle\, l\,\rangle_0$ such that $A_{\nu_r}(j_r, j_{r+1}) > 0$.

To see that (ii) follows, fix any $p \in \mathbb{Z}$ and define a sequence $(p_r)_{r=1}^{s+1}$ by setting $p_1 = p$ and $p_{r+1} = \nu_r + p_r - m$ for $r = 1, \dots, s$. Then $\nu_r = m + p_{r+1} - p_r$ and by definition each inequality $A_{\nu_r}(j_r, j_{r+1}) > 0$ ($r \in \langle s \rangle$) corresponds to an edge from (j_r, p_r) to (j_{r+1}, p_{r+1}) in the graph $G_m(A_0, A_1, \ldots, A_l)$, hence, we obtain a path $((j_r, p_r))_{r=1}^{s+1}$ of length s. Finally, (ii) follows by defining $q = p_{s+1}$.

To obtain (iii), analogously as done above, define a sequence $(p_r)_{r=1}^{s+1}$ by setting $p_{s+1} = q$ and $p_r = m + p_{r+1} - \nu_r$ for $r = s, \ldots, 1$. Then $\nu_r = m + p_{r+1} - p_r$ and similar to the proof of (ii), setting $p = p_1$, a path of length s to (k, q) is obtained starting from (j, p) with some level p.

Suppose that (ii) or (iii) holds. Then there exists a path $((j_r, p_r))_{r=1}^{s+1}$ in $G_m(A_0, A_1, \ldots, A_l)$ with $j_1 = j$ and $j_{s+1} = k$. Therefore, $A_{m+p_{r+1}-p_r}(j_r, j_{r+1}) > 0$, and thus $A(1)(j_r, j_{r-1}) > 0$ for $r = 1, \ldots, s$. Hence, $(j_r)_{r=1}^{s+1}$ is a path of length s from j to k in the associated directed graph of $A(1)$.

The last assertion follows from the constructions of the paths in this proof. □

An immediate consequence of this lemma is the following

Corollary 4.17. *Let P be an m-monic PFP of degree l. Then the following statements are equivalent.*

(i) *A is an irreducible matrix polynomial.*

(ii) *For all $j, k \in \langle n \rangle$ and $p \in \mathbb{Z}$ there exist a $q \in \mathbb{Z}$ and a path in $G_m(A_0, A_1, \ldots, , A_l)$ from (j, p) to (k, q),*

(iii) *For all $j, k \in \langle n \rangle$ and $q \in \mathbb{Z}$ there exist a $p \in \mathbb{Z}$ and a path in $G_m(A_0, A_1, \ldots, , A_l)$ from (j, p) to (k, q).*

We call a path $(j_1, p_1), \ldots, (j_{s+1}, p_{s+1})$ in $G_m(A_0, A_1, \ldots, , A_l)$ with $j_1 = j_{s+1}$ a **phase cycle** of $G_m(A_0, A_1, \ldots, , A_l)$ through $j_1 = j_{s+1}$. If (j, p) is a vertex of this phase cycle then there exists a phase cycle through j with same level displacement $p_{s+1} - p_1$, for example $(j_2, p_2), \ldots, (j_{s+2}, p_{s+2})$ with $j_{s+2} = j_2$ and $p_{s+2} = p_{s+1} + p_2 - p_1$ is a phase cycle through (j_2, p_2).

There are examples of graphs $G_m(A_0, A_1, \ldots, , A_l)$ such that the level displacements of all of its phase cycles are 0, see [FN05a, Example 4.3]

The **index of phase imprimitivity** of $G_m(A_0, \ldots, A_l)$ is defined as the g.c.d. (greatest common divisor) of the level displacements of all phase cycles in G. If the level displacements of all phase cycles in $G_m(A_0, \ldots, A_l)$ are zero, its index of phase imprimitivity is 0, by definition. Note that the index of phase imprimitivity of $G_m(A_0, \ldots, A_l)$ is a nonnegative integer.

Example 4.18. Consider

$$P(\lambda) = \lambda^2 I_3 - A(\lambda) = \lambda^2 I_3 - \begin{bmatrix} 0 & \lambda^3 & 0 \\ \lambda^2+\lambda & 0 & \lambda^2 \\ 0 & \lambda & 0 \end{bmatrix}.$$

A is irreducible and looking at the graph $G_2(A_0, A_1, A_2, A_3)$ of A shows that the index of imprimitivity is 1.

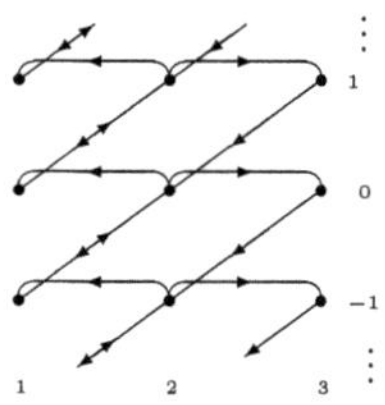

Figure 3: $G_2(A_0, A_1, A_2, A_3)$

Lemma 4.19. *Let P be an m-monic PFP of degree l with irreducible A and let d be the index of phase imprimitivity of the graph $G_m(A_0, \ldots, A_l)$. If $A_\mu(j,k) > 0$ and $A_\nu(j,k) > 0$ for some $j, k \in \langle n \rangle$ and some $\mu, \nu \in \langle l \rangle_0$, then d divides $\nu - \mu$. If $d = 0$, then $\mu = \nu$.*

Proof. For any $p \in \mathbb{Z}$ define $q := \mu - m + p$ and $u := \nu - m + p$. Then $((j,p),(k,q))$ and $((j,p),(k,u))$ are edges in $G_m = G_m(A_0, \ldots, A_l)$. Since A is irreducible, by Corollary 4.17 there exists a path in G_m from (k,q) to (j,v) for some $v \in \mathbb{Z}$. Then we have a phase cycle through j in G_m with level displacement $v - p$. Translating that path from (k,q) to (j,v) in G_m by changing the levels of its vertices by $\nu - \mu = u - q$ we obtain a path from (k,u) to $(j, v + \nu - \mu)$ and also a phase cycle trough j with level displacement $v + \nu - \mu - p$. Now d divides $v - p$ and $v + \nu - \mu - p$ and therefore it divides $\nu - \mu$. The last assertion follows immediately, since the level displacement of all phase cycles is 0 if $d = 0$. □

Lemma 4.20. *Let P be an m-monic PFP of degree l with irreducible A and let d be the index of phase imprimitivity of the graph $G_m(A_0, \ldots, A_l)$. For $j \in \langle n \rangle$ let d_j be the g.c.d. of the level displacements of all phase cycles of $G_m(A_0, \ldots, A_l)$ through phase j. Then $d_j = d$ and the set of the level displacements of all phase cycles of G through j is closed under addition.*

Proof. Clearly, $d \leqslant d_j$. We show that d_j is a common divisor of the level displacements of all phase cycles of G. Then the maximality of d implies the first assertion.

Let C be any phase cycle in $G = G_m(A_0, \ldots, A_l)$ and fix some vertex (i,p), $i \in \langle n \rangle$, $p \in \mathbb{Z}$ on C. Since C is a phase cycle in G, there is some $q \in \mathbb{Z}$ such that C is a path in G from (i,p) to (i,q), and hence, its level displacement is $q - p$.

By Corollary 4.17 there exists some $\mu \in \mathbb{Z}$, a path C_1 in G from (j,μ) to (i,p), a $\nu \in \mathbb{Z}$ and a path C_2 in G from (i,q) to (j,ν). Then the path $\hat{C}$ from (j,μ) to (j,ν) obtained by successively following the paths C_1, C and C_2 in the given order is a phase cycle in G through j with level displacement $\nu - \mu$.

Translate the path C_1 from (j,μ) to (i,p) by $q - p$ such that one obtains a path C_3 in G from $(j, \mu + q - p)$ to (i,q). Then the path C_0 from $(j, \mu + q - p)$ to (j,ν) obtained by successively passing along C_3 and C_2 is a phase cycle in G through j with level displacement $\nu - \mu - (q - p)$. Since d_j divides the level displacement $\nu - \mu$ of $\hat{C}$ and the level displacement $\nu - \mu - (q - p)$ of C_0, it also divides the level displacement $q-p$ of C. C was arbitrary, so d_j divides the level displacements of all phase cycles in G, hence, $d_j = d$.

By Remark 4.15 (i), for two cycles in G through j with level displacements k_1 and k_2 there is always a cycle in G through j with level displacement $k_1 + k_2$. This implies the second assertion. □

For the next important theorem we make use of the following technical lemma.

Lemma 4.21. *Let P be an m-monic PFP of degree l with irreducible A. Suppose that* $\operatorname{spr} A(\rho) = \rho^m$ *for some $\rho > 0$ and that $u \in \mathbb{R}^n$, $u > 0$ is an associated positive eigenvector of $A(\rho)$. Suppose, furthermore, that there exist a nonzero* $x = \begin{bmatrix} x_1 \\ \vdots \\ x_n \end{bmatrix} \in \mathbb{C}^n$ *and $\omega \in \mathbb{T}$ such that $P(\rho\omega)x = 0$. Then*

(i) for $k \in \langle l \rangle_0$ and $\mu, \tau \in \langle n \rangle$ we have that $A_k(\mu,\tau) \neq 0$ implies

$$\frac{x_\mu}{x_\tau} = \omega^{k-m}\frac{u_\mu}{u_\tau}; \tag{4-68}$$

(ii) for the level displacement $\tilde{d}$ of any phase cycle in $G_m(A_0,\ldots,A_l)$ we have $\omega^{\tilde{d}} = 1$.

Proof. (i) From Proposition 4.13 it follows that $|x| = \alpha u$ for some $\alpha > 0$. Therefore, all components x_μ ($\mu \in \langle n \rangle$) of x are nonzero. From $(\rho\omega)^m x = A(\rho\omega)x$ it follows that

$$\frac{x_\mu}{u_\mu} = \sum_{\nu=1}^{n}\sum_{j=0}^{l}\left(\rho^{j-m}A_j(\mu,\nu)\frac{u_\nu}{u_\mu}\right)\left(\omega^{j-m}\frac{x_\nu}{u_\nu}\right).$$

Now $\rho^m u = A(\rho)u$ implies that for $\mu = 1,\ldots,n$

$$1 = \sum_{\nu=1}^{n}\sum_{j=0}^{l}\rho^{j-m}A_j(\mu,\nu)\frac{u_\nu}{u_\mu}.$$

Therefore, since $\rho^{j-m}A_j(\mu,\nu)\frac{u_\nu}{u_\mu} \geqslant 0$, $\dfrac{x_\mu}{u_\mu}$ is a convex combination of the complex numbers $\omega^{j-m}\dfrac{x_\nu}{u_\nu}$, $\nu = 1,\ldots n$, which all, including $\frac{x_\mu}{u_\mu}$, lie on the circle $\mathbb{T}_\alpha$ due to

$$\left|\omega^{j-m}\frac{x_\nu}{u_\nu}\right| = \left|\frac{x_\nu}{u_\nu}\right| = \alpha = \left|\frac{x_\mu}{u_\mu}\right| \quad \text{for} \quad \nu = 1,\ldots,n.$$

Hence, $\frac{x_\mu}{u_\mu}$ is an extremal point of the convex hull of $\left\{\omega^{j-m}\frac{x_\nu}{u_\nu} : \nu \in \langle n \rangle\right\}$. This implies that for $j \neq k$, $\nu \neq \tau$ the coefficients $\rho^{j-m}A_j(\mu,\nu)\frac{u_\nu}{u_\mu}$ are zero and $\rho^{k-m}A_k(\mu,\tau)\frac{u_\tau}{u_\mu} = 1$. Hence, $\frac{x_\mu}{u_\mu} = \omega^{k-m}\frac{x_\tau}{u_\tau}$, which implies the assertion.

(ii) Let $C = ((\nu_k, p_k))_{k=1}^{s+1}$ be a phase cycle in the graph $G_m(A_0,\ldots,A_l)$. Hence, there exist some $j_{k_1}\ldots j_{k_s}$ such that $A_{j_k}(\nu_k,\nu_{k+1}) > 0$ and the level displacement of this phase cycle is $\tilde{d} = \sum\limits_{k=1}^{s}(j_k - m)$. Then, from (i) it follows that

$$\omega^{\tilde{d}} = \frac{x_{\nu_1}}{x_{\nu_{s+1}}}\prod_{k=1}^{s}\omega^{j_k-m}\frac{x_{\nu_{k+1}}}{x_{\nu_k}} = \frac{x_{\nu_1}}{x_{\nu_{s+1}}}\prod_{k=1}^{s}\frac{u_{\nu_{k+1}}}{u_{\nu_k}} = \frac{x_{\nu_1}}{x_{\nu_{s+1}}}\frac{u_{\nu_1}}{u_{\nu_{s+1}}} = 1,$$

where the last equality holds due to the fact that $\nu_1 = \nu_{s+1}$, since C is a phase cycle.

□

Due to [Min88], the following Lemma is a result of I. Schur. I thank Gabriele Penn-Karras for the main idea of the proof.

Lemma 4.22. *Let M be a nonempty set of integers which is closed under addition and let $\nu \in \mathbb{N}$ be the greatest common divisor* $\gcd(M)$ *of M. Then we have $\kappa\nu \in M$ for all but finitely many $\kappa \in \mathbb{N}$.*

Proof. If $\nu \in M$, then we have $M = \mathbb{N} \cdot \nu$, since M is closed under addition.

So let $\nu \notin M$. First note that there are finitely many integers $n_1, \dots, n_r \in M$ such that $\nu = \gcd(n_1, \dots, n_r)$. Indeed, if M is finite, this is clear. So suppose that $M = \{n_1, n_2, \dots\}$ and define the sequence $(g_j)_{j\in\mathbb{N}}$ via

$$g_1 = n_1 \quad \text{and} \quad g_{j+1} = \gcd(g_j, n_{j+1}).$$

Then $(g_j)_{j\in\mathbb{N}}$ is monotonically decreasing and, since it is bounded from below, there is an $r \in \mathbb{N}$ such that g_j remains constant for all $j \geqslant r$. This constant equals ν and therefore, $\gcd(M) = \gcd(n_1, \dots, n_r)$.

The set $M_r = \{n_1, \dots, n_r\}$ can be written as

$$M_r = \{k_1\nu, \dots, k_r\nu\},$$

where $k_1, \dots, k_r \in \mathbb{N}$ are certain integers. Consider the equation

$$\sum_{j=1}^{r} \alpha_j n_j = \sum_{j=1}^{r} \alpha_j k_j \nu = \nu$$

which is equivalent to

$$\sum_{j=1}^{r} \alpha_j k_j = 1. \tag{4-69}$$

By the theory of Diophantine equations (i. e. polynomial equations, investigated for integer or rational solutions; see e.g. [Mor69], [Ste05]), this equation has integer solutions $\alpha_1, \dots, \alpha_r \in \mathbb{Z}$ if and only if the right hand side of this equation is a multiple of $\gcd(k_1, \dots, k_r)$. Since $\nu = \gcd(n_1, \dots, n_r)$, it follows that $\gcd(k_1, \dots, k_r) = 1$. Otherwise, there would exist an integer which is larger than ν and divides all $n_1, \dots, n_r$. Hence, there exist $\alpha_1, \dots \alpha_r \in \mathbb{Z}$ such that (4-69) is satisfied. In other words, the g.c.d. ν of the set M is a linear combination of finitely many elements of M with integer coefficients.

Suppose w.l.o.g. that for $p \leqslant r$ the coefficients $\alpha_1, \dots, \alpha_p$ are positive, for $s \leqslant r - p$ the coefficients $\alpha_{p+1}, \dots, \alpha_{p+s}$ are negative and the remaining $r - p - s$ coefficients $\alpha_{p+s+1}, \dots, \alpha_r$ are zero. Then, setting $\tilde{\alpha}_j = -\alpha_{p+j} > 0$ for $j \in \langle\, s \,\rangle$, we obtain

$$\nu = \sum_{j=1}^{p} \alpha_j n_j - \sum_{j=1}^{s} \tilde{\alpha}_j n_{p+j} =: \mu_p - \mu_s.$$

Since $\nu > 0$, it follows that $p > 0$ and, since $\alpha_1, \dots, \alpha_p \in \mathbb{Z}$ and since M is closed under addition, we have that $\mu_p \in M$. But then $\nu \notin M$ implies that $s > 0$, hence, also $\mu_s \in M$. Furthermore, $\nu > 0$ implies that $\mu_p > \mu_s$. Therefore,

$$\gcd(\mu_p, \mu_s) = \gcd(\mu_p, \mu_p - \mu_s) = \gcd(\mu_s, \nu) = \nu.$$

Define $\kappa_0 = \frac{\mu_p \mu_s}{\nu^2}$. We now show that for all $\kappa \geqslant \kappa_0$ we have that $\kappa\nu \in M$.

For $x \in \mathbb{R}$ define $\lceil x \rceil = \min_{n \in \mathbb{Z}, n \geqslant x} n$. Let $\kappa \geqslant \kappa_0$ and define the integers

$$a = \kappa - \frac{\mu_s}{\nu} \left\lceil \frac{\kappa\nu}{\mu_p} \right\rceil \quad \text{and} \quad b = \frac{\mu_p}{\nu} \left\lceil \frac{\kappa\nu}{\mu_p} \right\rceil - \kappa$$

Obviously, $b \geqslant 0$. Furthermore, by definition of κ_0, we have

$$\kappa\nu\mu_p - \kappa\nu\mu_s = \kappa\nu(\mu_p - \mu_s) = \kappa\nu^2 \geqslant \mu_p\mu_s.$$

This implies that $\frac{\kappa\nu}{\mu_s} - 1 \geqslant \frac{\kappa\nu}{\mu_p}$, hence, $\frac{\kappa\nu}{\mu_s} \geqslant \left\lceil \frac{\kappa\nu}{\mu_p} \right\rceil$, i. e., $a \geqslant 0$.

Since M is closed under addition, it follows that $0 < \kappa\nu = a\mu_p + b\mu_s \in M$.

□

Theorem 4.23. *Let P be an m-monic PFP of degree l with irreducible A. Let d be the index of phase imprimitivity of the graph $G = G_m(A_0, A_1, \dots, A_l)$. Then for all $\rho > 0$ with $\operatorname{spr} A(\rho) = \rho^m$ the following statements hold.*

(*i*) *$d = 0$ is equivalent to $\sigma(P) = \mathbb{C}$, i. e., P is singular.*

(*ii*) *Let $d > 0$. Then for $\theta \in [0, 2\pi)$*

$$\rho e^{i\theta} \in \sigma(P) \quad \textit{if and only if} \quad \theta \in \left\{0, \frac{2\pi}{d}, 2 \cdot \frac{2\pi}{d}, \dots, (d-1)\frac{2\pi}{d}\right\}.$$

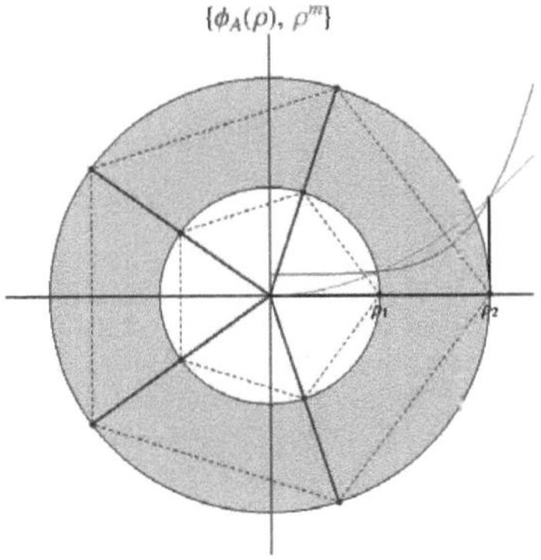

Figure 4: example for $d = 5$

Proof. We will start with an observation that will be useful in both the proof of (*i*) and (*ii*).

(*o*) Let $\omega \in \mathbb{T}$ be such that $\omega^d = 1$. Note that we can choose any $\omega \in \mathbb{T}$ if $d = 0$. Choose a strictly positive eigenvector $u = \begin{bmatrix} u_1 \\ \vdots \\ u_n \end{bmatrix} \in \mathbb{C}^n$ of P to its eigenvalue ρ. Set $x_1 = u_1$ and

$$\omega^{-j+m} \frac{u_\nu}{u_\mu} x_\mu = x_\nu \quad \text{if} \quad A_j(\mu, \nu) \neq 0,$$

for $\mu \in \langle n \rangle$. The numbers x_ν are are well defined, i. e., they do not depend on j. Indeed, if $x_\nu^{(1)} = \omega^{-j_1+m} \frac{u_\nu}{u_\mu} x_\mu$ and $x_\nu^{(2)} = \omega^{-j_2+m} \frac{u_\nu}{u_\mu} x_\mu$, then $\frac{x_\nu^{(1)}}{x_\nu^{(2)}} = \omega^{j_2-j_1}$ and by Lemma 4.19 $j_2 - j_1 = kd$ for some $k \in \mathbb{Z}$. Hence, since $\omega^d = 1$, $x_\mu^{(1)} = x_\mu^{(2)}$. Furthermore, due to the irreducibility of A, for each $\nu \in \langle n \rangle$, x_ν is assigned a value. Set $x = \begin{bmatrix} x_1 \\ \vdots \\ x_n \end{bmatrix}$.

By Lemma 4.21, we have for $\mu \in \langle n \rangle$

$$\begin{aligned}
\big(A(\omega\rho)x\big)_\mu &= x_\mu \sum_{\nu=1}^{n} \sum_{j=0}^{l} A_j(\mu,\nu)(\omega\rho)^j \frac{x_\nu}{x_\mu} = x_\mu \sum_{\nu=1}^{n} \sum_{j=0}^{l} A_j(\mu,\nu)\rho^j \omega^m \frac{u_\nu}{u_\mu} \\
&= \omega^m \frac{x_\mu}{u_\mu} \sum_{j=0}^{l} \rho^j \sum_{\nu=1}^{n} A_j(\mu,\nu) u_\nu = \omega^m \frac{x_\mu}{u_\mu} \sum_{j=0}^{l} \rho^j (A_j u)_\mu \\
&= \omega^m \frac{x_\mu}{u_\mu} \big(A(\rho)u\big)_\mu = \omega^m \frac{x_\mu}{u_\mu} \rho^m u_\mu = \omega^m \rho^m x_\mu,
\end{aligned}$$

hence, $\rho^m \omega^m x = A(\omega\rho)x$. Therefore, $\omega\rho$ is an eigenvalue of P and x is a corresponding eigenvector.

(*i*) If $\mathbb{T}_\rho \subset \sigma(P)$, then Lemma 4.21 implies that $\omega^{\tilde{d}} = 1$ for all $\omega \in \mathbb{T}$ for the level displacement $\tilde{d}$ of any phase cycle of $G = G_m(A_0, \dots, A_l)$. Hence, $\tilde{d} = 0$. Therefore, the level displacements of all phase cycles in G are zero, i. e. $d = 0$.

Conversely, if $d = 0$ then by (*o*) we obtain $\mathbb{T}_\rho \subset \sigma(P)$.

Hence, $d = 0$ if and only if $\mathbb{T}_\rho \subset \sigma(P)$, i. e., $\sigma(P) = \mathbb{C}$.

(*ii*) Let $d > 0$. (*o*) shows that

$$\left\{\rho\, e^{i\varphi} \colon \varphi = 0, \frac{2\pi}{d}, \frac{4\pi}{d}, \quad \dots \quad , (d-1)\frac{2\pi}{d}\right\} \subset \sigma(P).$$

Suppose that $\rho\, e^{i\theta}$ is an eigenvalue of P. From Lemma 4.21(*ii*), it follows that for any level displacement $\tilde{d}_j$ of any phase cycle in G through some phase j, $\tilde{d}_j\, \theta$ is a multiple of 2π. Due to Lemma 4.20, d is the g.c.d. of all level displacements $\tilde{d}_j$ of all phase cycles through j. From Lemma 4.20 and Lemma 4.22 we know that all but a finite number of multiples of d are level displacements of some phase cycles in G through j. Therefore, there exists a $k \in \mathbb{Z}$ such that $kd\theta$ and $(k+1)d\theta$ are multiples of 2π and hence, $d\theta$ is a multiple of 2π. Thus, $\theta \in \{k\frac{2\pi}{d} \colon k \in \mathbb{Z}\}$, i. e.

$$\rho\, e^{i\theta} \in \left\{\rho e^{i\varphi} \colon \varphi = 0, \frac{2\pi}{d}, \frac{4\pi}{d}, \dots, (d-1)\frac{2\pi}{d}\right\}.$$

□

Remark 4.24. *Notice that the angles of rotation invariance in Theorem 4.23 (ii) only depend on the index of phase imprimitivity of the graph G. This implies that if there is more than one number $\rho > 0$ that satisfies $\phi_A(\rho) = \rho^m$, then the angles of rotation invariance are the same for each of these numbers.*

Proposition 4.25. *Let P be an m-monic PFP of degree l with irreducible A. Then the following statements hold.*

(i) *If P has a right PFF C, i. e.,*

$$P(\lambda) = B(\lambda)C(\lambda) = \left(\sum_{j=0}^{l-m} \lambda^j B_j\right)\left(\lambda^m I_n - \sum_{j=0}^{m-1} \lambda^j C_j\right) \tag{4-70}$$

with $C_j \geqslant 0$ $(j \in \langle m-1\rangle_0)$, and if $\rho := \operatorname{spr}(C) > 0$, then

(a) $\phi_A(\rho) = \operatorname{spr} A(\rho) = \rho^m$.

(b) $\mathsf{N}(C(\rho)) = \mathsf{N}(P(\rho)) = \operatorname{span}\{v\}$ *for some* $v > 0$.

(c) *C has algebraically simple peripheral eigenvalues.*

(ii) *If there is a number $\tau > 0$ such that $\phi_A(\tau) = \operatorname{spr} A(\tau) = \tau^m$, then P has a right PFF $C(\lambda) = \lambda^m I_n - \sum_{j=0}^{m-1} \lambda^j C_j$ with $\operatorname{spr}(C) \leqslant \tau$.*

(iii) *If P has a right PFF and if the left divisor $B(\lambda)$ in (4-70) satisfies $\operatorname{spr}(\operatorname{rev} B) > 0$ and if B_0 is a nonsingular M-matrix, then $\operatorname{rev} B$ has algebraically simple peripheral eigenvalues.*

(iv) *Let $\widetilde{C}$ be the minimal PFF of degree m of P with $\operatorname{spr}(\widetilde{C}) > 0$. Then*

$$\operatorname{spr} \widetilde{C} = \min\{\rho > 0 \mid \operatorname{spr} A(\rho) = \rho^m\} := \tilde{\rho},$$
$$\sigma(\widetilde{C}) = \sigma(P) \cap \overline{\mathbb{D}}_{\tilde{\rho}}.$$

Proof. (i) (a) By Corollary 3.7, $\operatorname{spr}(C)$ is an eigenvalue of C and thus also for P. Hence, there exists an entrywise nonnegative vector $v \neq 0$ such that

$$\rho^m v = A(\rho)v.$$

Since $A(\rho)$ is an irreducible matrix, v must be strictly positive (see e.g. [Min88, Theorem I.2.2, p.7]). Hence, $\rho^m = \operatorname{spr} A(\rho)$ (see e.g. [Min88, Theorem I.4.4, p 16]), which proves (a).

(b) ρ is an eigenvalue of C and therefore, it is an eigenvalue of P. $A(\rho)$ is irreducible, so ρ^m is an algebraically and hence, geometrically simple eigenvalue of $A(\rho)$. Let v be a strictly positive eigenvector v of $A(\rho)$ corresponding to ρ^m. Note that $\mathsf{N}(C(\rho)) \neq \{0\}$, since ρ is an eigenvalue of C. Therefore

$$\{0\} \neq \mathsf{N}(C(\rho)) \subset \mathsf{N}(P(\rho)) = \mathsf{N}\big(\rho^m I_n - A(\rho)\big) = \operatorname{span}\{v\}$$

implies that $\mathsf{N}(C(\rho)) = \mathsf{N}(P(\rho)) = \operatorname{span}\{v\}$.

(c) Let $\lambda \in \mathbb{C}$ be an eigenvalue of C with $|\lambda| = \rho$ and let $w \neq 0$ be a corresponding eigenvector. Then $P(\lambda)w = 0$, i. e. $\lambda^m w = A(\lambda)w$. Taking absolute values componentwise on both sides leads to $\rho^m |w| \leqslant A(\rho)|w|$. Let $u > 0$ be a left eigenvector of $A(\rho)$ corresponding to ρ^m, i. e., $A(\rho)^T u = \rho^m u$. Then, since $u > 0$,

$$\rho^m \langle |w|, u\rangle \leqslant \langle A(\rho)|w|, u\rangle = \langle |w|, A(\rho)^T u\rangle = \rho^m \langle |w|, u\rangle . \qquad (4\text{-}71)$$

Thus $\langle (\rho^m I_n - A(\rho))|w|, u\rangle = 0$ and, since $u > 0$, it follows that $(\rho^m I_n - A(\rho))|w| = 0$. Since $A(\rho)$ is irreducible and ρ^m is its spectral radius, $|w|$ is a multiple of v.

Now, take any two vectors $w_1, w_2 \in \mathbb{C}^n$ with $w_1, w_2 \in \mathsf{N}(\lambda^m I_n - A(\lambda))$ and $\alpha \in \mathbb{C}$ such that the first component of the vector $\widetilde{w} = \alpha w_1 + w_2$ is zero. Then analogously to (4-71) it follows that $(\rho^m I_n - A(\rho))|\widetilde{w}| = 0$. Since $|\widetilde{w}|$ is not strictly positive, it follows that $\widetilde{w} = 0$ and therefore w_1, w_2 are linearly dependent. This proves that λ is a geometrically simple eigenvalue of $A(\lambda)$, i. e., $\dim \mathsf{N}(P(\lambda)) = 1$. With $\{0\} \neq \mathsf{N}(C(\lambda)) \subset \mathsf{N}(P(\lambda))$ the second part of the assertion follows, see [Sch74, Ex.8(a), p.43], [Sch86, Cor. 3.5].

(ii) Since $A(\tau)$ is an irreducible matrix, there exists a positive vector $u > 0$ such that $A(\tau)u = \tau^m u$, hence, u is an eigenvector of P to the eigenvalue τ. By Proposition 3.9 the positive vector $\hat{u} = \Lambda(\tau)u = \begin{bmatrix} \tau^{m-1}u & \cdots & \tau u & u \end{bmatrix}^T$ is an eigenvector of $\mathcal{P}$ to the eigenvalue $\tau > 0$, where $\mathcal{P}(\lambda) = \lambda I_n - \mathcal{A}(\lambda) = \lambda I_n - \sum_{j=0}^{l-m+1} \lambda^j \mathcal{A}_j$ is the canonical reduction of P. So $\mathcal{P}$ is a 1–monic matrix polynomial with nonnegative coefficients $\mathcal{A}_0, \ldots, \mathcal{A}_{l-m+1}$ and with $\mathcal{A}(\tau)\hat{u} = \tau\hat{u}$, $\hat{u} > 0$. Therefore, by Proposition 4.5, $\mathcal{P}$ has a nonnegative right root $\mathcal{C}$ with $\mathrm{spr}(\mathcal{C}) \leqslant \tau$. Corollary 3.6 then implies that P has a factorization

$$P(\lambda) = B(\lambda)C(\lambda) = \left(\sum_{j=0}^{l-m} \lambda^j B_j\right)\left(\lambda^m I_n - \sum_{j=0}^{m-1} \lambda^j C_j\right)$$

with $\mathrm{spr}(C) \leqslant \tau$. Since $\mathcal{C}$ is entrywise nonnegative and it is the companion matrix of C, the coefficients $C_0, \ldots, C_{m-1}$ are nonnegative.

(iii) Since by assumption 0 is not an eigenvalue of B, the coefficient B_0 is nonsingular. Thus setting $\widetilde{B}_j = -B_j B_0^{-1}$, we have

$$\mathrm{rev}\, B(\lambda) = \left(\lambda^{l-m} I_n - \sum_{j=1}^{l-m} \lambda^{l-m-j} \widetilde{B}_j\right) B_0 = \mathrm{rev}\, \widetilde{B}(\lambda) B_0.$$

B_0 is a nonsingular M–matrix, so $\widetilde{B}_j = B_j B_0^{-1} \geqslant 0$ for all $j \in \langle\, l-m\,\rangle$ and

$$\mathrm{rev}\, P(\lambda) = \lambda^l P(1/\lambda) = \lambda^{l-m} \widetilde{B}(1/\lambda)\, B_0\, \lambda^m C(1/\lambda) = \mathrm{rev}\, \widetilde{B}(\lambda) B_0\, \mathrm{rev}\, C(\lambda),$$

i. e., $\mathrm{rev}\, \widetilde{B}^T$ is a right PFF of the $(l-m)$–monic matrix polynomial $\mathrm{rev}\, P^T(\lambda) = \lambda^{l-m} I_n - \sum_{j=0}^{l} \lambda^{l-j} A_j^T$. Since $\mathrm{spr}(\mathrm{rev}\, \widetilde{B}) > 0$, we can apply (iii) to $\mathrm{rev}\, \widetilde{B}^T$, so $\mathrm{rev}\, \widetilde{B}^T$ and therefore $\mathrm{rev}\, B = (\mathrm{rev}\, \widetilde{B})B_0$ has simple peripheral eigenvalues.

(*iv*) Suppose that $\tilde{\mathcal{C}}$ is the companion matrix of the minimal PFF $\tilde{C}$. Then we have $\tilde{\mathcal{C}} \leqslant \mathcal{C}$ for the companion matrix $\mathcal{C}$ of any other PFF C of P. $\tilde{\mathcal{C}}$ and $\mathcal{C}$ are entrywise nonnegative, hence, $\mathrm{spr}(\tilde{C}) = \mathrm{spr}(\tilde{\mathcal{C}}) \leqslant \mathrm{spr}(\mathcal{C}) = \mathrm{spr}(C)$. The first equality then follows from (*i*) of this proposition.

For the second identity, note that clearly $\sigma(\widetilde{C}) \subset \sigma(P) \cap \overline{\mathbb{D}}_{\tilde{\rho}}$. The remaining inclusion follows from Proposition 1.15.

□

Following up Proposition 1.21 we formulate the next lemma.

Lemma 4.26. *Let P be an m-monic PFP of degree l with irreducible A.*

(*i*) *Let* $\mathrm{spr}(A(\rho)) = \rho^m > 0$. *Then the following relations hold.*

$$\begin{aligned}(\det P)'(\rho) = 0 \quad &\textit{if and only if} \quad (\mathrm{spr}\, A)'(\rho) = m\rho^{m-1},\\ (\det P)'(\rho) > 0 \quad &\textit{if and only if} \quad (\mathrm{spr}\, A)'(\rho) < m\rho^{m-1} \textit{ and}\\ (\det P)'(\rho) < 0 \quad &\textit{if and only if} \quad (\mathrm{spr}\, A)'(\rho) > m\rho^{m-1}.\end{aligned}$$

(*ii*) *Let* $(\mathrm{spr}\, A)(\rho) = \rho^m > 0$ *and* $(\mathrm{spr}\, A)'(\rho) = m\rho^{m-1}$. *Then*

$$(\det P)''(\rho) \neq 0 \quad \textit{if and only if} \quad (\mathrm{spr}\, A)''(\rho) \neq m(m-1)\rho^{m-2}.$$

Proof. Since A is irreducible, the function ϕ_A is differentiable for all positive τ, see the introduction of this section. By Proposition 1.12 and Proposition 1.21

$$(\det P)'(\rho) = \left(m\rho^{m-1} - \phi_A'(\rho)\right) \mathrm{tr}[\mathrm{adj}(P(\rho))]$$

and

$$(\det P)''(\rho) = \left(m(m-1)\rho^{m-2} - \phi_A''(\rho)\right) \mathrm{tr}[\mathrm{adj}(P(\rho))].$$

Using [Min88, Corollary 4.1, p.16] we obtain $\mathrm{adj}(P(\rho)) > 0$, hence, we have $\mathrm{tr}[\mathrm{adj}(P(\rho))] > 0$. The assertions (*i*) and (*ii*) then follow immediately, (see also [GHT96] and [GHT98, p.544]).

□

Proposition 4.27. *Let P be an m-monic PFP of degree l with irreducible A. Suppose further, that $\phi_A(\rho) = \mathrm{spr}\, A(\rho) = \rho^m$ for some $\rho > 0$. Then the following statements hold.*

(*i*) *If $\phi_A'(\rho) \neq m\rho^{m-1}$, then the eigenvalues of P on $\mathbb{T}_\rho$ are simple; i. e., their geometric and algebraic multiplicities are 1.*

(*ii*) *If $\phi_A'(\rho) = m\rho^{m-1}$ then either $\mathrm{spr}\, A(\tau) = \tau^m$ for all $\tau \geqslant 0$ or $\mathrm{spr}\, A(\tau) > \tau^m$ for all positive $\tau \neq \rho$.*

In the second case we have $\phi_A''(\rho) > m(m-1)\rho^{m-2}$, and the eigenvalues of P on $\mathbb{T}_\rho$ have geometric multiplicity 1 and algebraic multiplicity 2.

Proof. (*i*) Suppose that $\phi_A'(\rho) < m\rho^{m-1}$. Then there is a $\delta > 0$ such that $\phi_A(\tau) < \tau^m$ for all $\tau \in (\rho, \rho + \delta)$. By Proposition 1.15, P has exactly nm eigenvalues (counting multiplicities) in $\overline{\mathbb{D}}_\rho$. Fix one $\tau \in (\rho, \rho+\delta)$. By Theorem 4.1, P has a nonnegative spectral factorization

$$P(\lambda) = \widetilde{B}(\lambda)B_0C(\lambda) \tag{4-72}$$

with respect to the circle $\mathbb{T}_\tau$ where C is a monic matrix polynomial. Since $\deg C = m$, C has nm eigenvalues and due to $\sigma(C) = \sigma(P) \cap \mathbb{D}_\tau$ they lie in $\mathbb{D}_\tau \supset \overline{\mathbb{D}}_\rho$. There are already nm eigenvalues of C in $\overline{\mathbb{D}}_\rho$, thus $\operatorname{spr} C \leqslant \rho$. Now $\rho \in \sigma(P)$ implies $\operatorname{spr} C = \rho$. By Proposition 4.25 the eigenvalues of C on $\mathbb{T}_\rho$ are simple. Since $\sigma(\widetilde{B}) = \sigma(P) \cap (\mathbb{C} \setminus \overline{\mathbb{D}}_\tau)$, B does not have eigenvalues on $\mathbb{T}_\rho$, hence, the assertion follows.

Now suppose that $\phi_A'(\rho) > m\rho^{m-1}$. Then there exists a $\delta > 0$ such that $\phi_A(\tau) < \tau^m$ for all $\tau \in (\rho-\delta, \rho)$. Fix $\tau \in (\rho-\delta,\rho)$. By Theorem 4.1, P has a spectral nonnegative factorization (4-72) with respect to $\mathbb{T}_\tau$ where $\widetilde{B}$ is a comonic matrix polynomial with $\sigma(\widetilde{B}) = \sigma(P) \cap (\mathbb{C} \setminus \overline{\mathbb{D}}_\tau)$ and therefore $\operatorname{spr}(\operatorname{rev}\widetilde{B}) < \frac{1}{\tau}$. By Proposition 1.15 *(iii)*, P has no eigenvalues in $\mathbb{A}_{\rho-\delta,\rho}$ and therefore, since $\sigma(C) = \sigma(P) \cap \mathbb{D}_\tau$, also C does not have any eigenvalues there. Hence, $\operatorname{spr} C < \rho$. Since by assumption ρ is an eigenvalue of P, it is an eigenvalue of $\widetilde{B}$ and therefore, $\operatorname{spr}(\operatorname{rev}\widetilde{B}) = \frac{1}{\rho} > 0$. Proposition 4.25 *(iii)* then implies that $\operatorname{rev}\widetilde{B}$ has simple peripheral eigenvalues, hence, the eigenvalues of $\widetilde{B}$ on $\mathbb{T}_\rho$, and therefore the eigenvalues of P on $\mathbb{T}_\rho$ are simple.

(ii) Suppose that $\phi_A(\tau) = \tau^m$ does not hold for all $\tau \geqslant 0$. Consider the function $\eta_A : \mathbb{R} \to \mathbb{R}$ with $\eta_A(t) = \ln \phi_A(e^t)$. By the geometric convexity of ϕ_A, η_A is convex in t. Set $r = \ln\rho$. Then $\eta_A(r) = mr$ and further we have that $\eta_A(t) = mt$ does not hold for all $t \in \mathbb{R}$. Thus $\eta_A(t) > mt$ for all $t \in \mathbb{R}$, $t \neq r$, i. e., $\phi_A(\tau) > \tau^m$ for all $\tau \geqslant 0$, $\tau \neq \rho$. Thus the first statement of *(ii)* follows.

Suppose now that $\phi_A(\tau) > \tau^m$ for all $\tau > 0$, $\tau \neq \rho$. For $\epsilon \in (0,1)$ define $P_\epsilon(\lambda) := \lambda^m I_n - \epsilon A(\lambda)$. Then $\operatorname{spr}(\epsilon A(\rho)) < \rho^m$ and by Theorem 4.1 we have, therefore,

$$P_\epsilon(\lambda) = \widetilde{B}_\epsilon(\lambda)B_{0,\epsilon}C_\epsilon(\lambda),$$

where $C_\epsilon(\lambda) = \lambda^m I_n - \sum\limits_{j=0}^{m-1} \lambda^j C_{j,\epsilon}$ is the minimal right PFF of P_ϵ with

$$\sigma(C_\epsilon) = \sigma(P_\epsilon) \cap \mathbb{D}_\rho, \tag{4-73}$$

$\widetilde{B}_\epsilon(\lambda) = I_n - \sum\limits_{j=1}^{l-m} \lambda^j \widetilde{B}_{j,\epsilon}$ is a comonic matrix polynomial of degree $l-m$ with

$$\sigma(\widetilde{B}_\epsilon) = \sigma(P_\epsilon) \cap (\mathbb{C} \setminus \overline{\mathbb{D}}_\rho) \tag{4-74}$$

and $B_{0,\epsilon}$ is a nonsingular M-matrix. For $j \in \langle l-m \rangle$ set $B_{j,\epsilon} = -\widetilde{B}_{j,\epsilon}B_{0,\epsilon}$, thus,

$$P_\epsilon(\lambda) = \left(\sum_{j=0}^{l-m} \lambda^j B_{j,\epsilon} \right) C_\epsilon(\lambda).$$

Because $\phi_A(\rho) = \rho^m$, from Proposition 4.25 we know that P has a minimal right PFF C. Hence, there is a matrix polynomial $B(\lambda) = \sum_{j=0}^{l-m} \lambda^j B_j$ such that

$$P(\lambda) = B(\lambda)C(\lambda).$$

We will show that for all $j \in \langle l-m \rangle_0$ we have that $B_{j,\epsilon} \to B_j$ as $\epsilon \to 1$.

Consider the canonical reduction $\mathcal{P}_\epsilon$ of P_ϵ to degree $k = l - m + 1$ according to Corollary 3.3. Then $\mathcal{P}_\epsilon(\lambda) = \lambda I_{mn} - \mathcal{A}_\epsilon(\lambda)$, where $\mathcal{A}_\epsilon(\lambda) = \sum_{j=0}^{k} \lambda^j \mathcal{A}_{j,\epsilon}$ with

$$\mathcal{A}_{0,\epsilon} = \begin{bmatrix} \epsilon A_{m-1} & \cdots & \cdots & \epsilon A_0 \\ I_n & & & \vdots \\ & \ddots & & \vdots \\ & & I_n & 0 \end{bmatrix} \text{ and } \mathcal{A}_{j,\epsilon} = \begin{bmatrix} \epsilon A_{j+m-1} & & 0 \\ & \ddots & \\ & & 0 \end{bmatrix} \text{ for } j = 1, \ldots, k.$$

Let $\mathcal{C}_\epsilon$ and $\mathcal{C}$ be the companion matrices of C_ϵ and C, respectively. By Corollary 3.6, $\mathcal{C}_\epsilon$ is the minimal right root of $\mathcal{P}_\epsilon$, i. e.,

$$\mathcal{P}_\epsilon(\lambda) = \mathcal{B}_\epsilon(\lambda)(\lambda I_{mn} - \mathcal{C}_\epsilon),$$

where $\mathcal{B}_\epsilon(\lambda) = \sum_{j=0}^{l-m} \lambda^j \mathcal{B}_{j,\epsilon}$ and $\mathcal{C}$ is the minimal right root of $\mathcal{P}$, i. e.,

$$\mathcal{P}(\lambda) = \mathcal{B}(\lambda)(\lambda I_{mn} - \mathcal{C})$$

with $\mathcal{B}(\lambda) = \sum_{j=0}^{l-m} \lambda^j \mathcal{B}_j$.

Let $0 < \epsilon_1 \leqslant \epsilon_2 < 1$ and denote by $\left(\mathcal{C}_k^{(\epsilon_1)}\right)_{k\in\mathbb{N}}$ and $\left(\mathcal{C}_k^{(\epsilon_2)}\right)_{k\in\mathbb{N}}$ the sequences generated by the fixpoint iteration (4-57) with initial matrix 0, i.e,

$$\mathcal{C}_0^{(\epsilon_\nu)} = 0, \quad \mathcal{C}_{k+1}^{(\epsilon_\nu)} = \sum_{j=0}^{l} \mathcal{A}_{j,\epsilon_\nu} \left(\mathcal{C}_k^{(\epsilon_\nu)}\right)^j \quad \text{for } \nu = 1, 2$$

and let $\mathcal{C}_{\epsilon_1}$ and $\mathcal{C}_{\epsilon_2}$ be their limits, respectively. Then obviously $\mathcal{C}_{\epsilon_1} \leqslant \mathcal{C}_{\epsilon_2} \leqslant \mathcal{C}$ and hence, the (componentwise) limit of $\mathcal{C}_\epsilon$ exists as ϵ goes to 1. By Proposition 4.4 we have $\lim_{\epsilon\to 1} \mathcal{C}_\epsilon = \mathcal{C}$. From (4-53) for $j \in \langle l-m \rangle_0$ it follows that

$$\mathcal{B}_{j,\epsilon} = \delta_{0j} I_{mn} - \sum_{i=0}^{l-m-j} \epsilon \mathcal{A}_{j+i+1} \mathcal{C}_\epsilon^j, \quad \text{and} \quad \mathcal{B}_j = \delta_{0j} I_{mn} - \sum_{i=0}^{l-m-j} \mathcal{A}_{j+i+1} \mathcal{C}^j,$$

hence, also $\mathcal{B}_{j,\epsilon} \to \mathcal{B}_j$ as $\epsilon \to 1$.

Due to the form of $\mathcal{C}_\epsilon$ and $\mathcal{C}$, also $C_{j,\epsilon} \to C_j$ for $\epsilon \to 1$. From the Remark 3.5 it follows that the upper left $n \times n$ blocks of $\mathcal{B}_{j,\epsilon}$ and $\mathcal{B}_j$ coincide with $B_{j,\epsilon}$ and B_j, respectively, for $j \in \langle l-m \rangle_0$. Therefore, $B_{j,\epsilon} \to B_j$ as $\epsilon \to 1$ for $j \in \langle l-m \rangle_0$.

Therefore, since $\widetilde{B}_\epsilon$ has no eigenvalues in $\overline{\mathbb{D}}_\rho$ for all $\epsilon \in (0,1)$ and the eigenvalues of $\widetilde{B}_\epsilon$ depend continuously on ϵ, it follows that B has no eigenvalues in $\mathbb{D}_\rho$. Hence, 0 is not an eigenvalue of B and therefore, B_0 is nonsingular. Furthermore, due to Lemma 4.9, B_0 is a Z–matrix. Since $B_{0,\epsilon}^{-1} \geqslant 0$ for all $\epsilon \in (0,1)$ and $B_{0,\epsilon} \to B_0$ as $\epsilon \to 1$, it follows that also $B_0^{-1} \geqslant 0$, i. e., B_0 is a nonsingular M-matrix.

Therefore, the left factor B of P can be expressed as $B(\lambda) = \widetilde{B}(\lambda) B_0$ with a comonic matrix polynomial $\widetilde{B}(\lambda) = I_n - \sum_{j=1}^{l-m} \lambda^j \widetilde{B}_j$.

Due to $\phi_A'(\rho) = m\rho^{m-1}$ and $\phi_A(\tau) > \tau^m$ for all positive $\tau \neq \rho$, the geometrical convexity of ϕ_A implies that there is a $\delta < 1$ such that for all $\epsilon \in (\delta, 1)$ there are exactly two positive numbers $\tau_{\epsilon,1}, \tau_{\epsilon,2}$ with $\tau_{\epsilon,1} < \rho < \tau_{\epsilon,2}$ such that $\phi_{\epsilon A}(\tau_{\epsilon,\nu}) = \tau_{\epsilon,\nu}^m$ for $\nu = 1, 2$. Since the roots of the function

$$\tau \mapsto \operatorname{spr}(\epsilon A(\tau)) - \tau^m = \epsilon \operatorname{spr}(A(\tau)) - \tau^m, \quad \epsilon \in (\delta, 1)$$

depend continuously on ϵ, we have that $\tau_{\epsilon,1} \to \rho$ and $\tau_{\epsilon,2} \to \rho$ as $\epsilon \to 1$.

Let d denote the index of phase imprimitivity of $G_1(A_0, \ldots, A_k)$. By assumption and Remark 1.19, P is regular, hence, Theorem 4.23 implies that $d > 0$. Note that $G(A_0, \ldots, A_l) = G(\epsilon A_0, \ldots, \epsilon A_l)$. So by Theorem 4.23, $\tau_{\epsilon,1}e^{i\theta}$ and $\tau_{\epsilon,2}e^{i\theta}$ are eigenvalues of P_ϵ if and only if $\theta \in \{\nu\frac{2\pi}{d} : \nu \in \langle d-1 \rangle_0\}$.

Remember that due to Proposition 1.7, P has no eigenvalues in the annulus $\mathbb{A}_{\tau_{\epsilon,1},\tau_{\epsilon,2}}$. This together with (4-73) and (4-74) now implies that $\operatorname{spr}(C_\epsilon) = \tau_{\epsilon,1}$ and $\operatorname{spr}(\operatorname{rev} B_\epsilon) = \tau_{\epsilon,2}$. Letting $\epsilon \to 1$, we obtain that $\operatorname{spr}(C) = \operatorname{spr}(\operatorname{rev} \widetilde{B}) = \rho$ and $\{\rho e^{i\theta} : \theta = \nu\frac{2\pi}{d}, \nu \in \langle d-1 \rangle_0\}$ are the peripheral eigenvalues of both C and $\operatorname{rev} \widetilde{B}$. By Proposition 4.25 they are simple eigenvalues for both, C and $\operatorname{rev} \widetilde{B}$ and hence, also for $\widetilde{B}$. Since $\det P(\lambda) = \det \widetilde{B}(\lambda) \det B_0 \det C(\lambda)$, their algebraic multiplicities as eigenvalues of P are 2. By Proposition 4.13, they are of geometric multiplicity 1.

To verify the remaining assertion, note that, since ρ is an algebraically double eigenvalue of P, $(\det P)''(\rho) \neq 0$ and by Lemma 4.26 we have $\phi_A''(\rho) \neq m(m-1)\rho^{m-2}$. Since ϕ_A is geometrically convex, the function $\eta : t \mapsto \ln(\phi_A(e^t))$ is convex. Let t_0 be such that $\rho = e^{t_0}$. Then $\eta''(t_0) = 0$ if and only if

$$\frac{\phi_A''(\rho)\phi_A(\rho) - \phi_A'(\rho)^2}{\phi_A(\rho)^2}\rho^2 + \frac{\phi_A'(\rho)}{\phi_A(\rho)}\rho = 0. \tag{4-75}$$

Using $\phi_A(\rho) = \rho^m$ and $\phi_A'(\rho) = m\rho^{m-1}$ the last identity is equivalent to

$$\phi_A''(\rho) = m(m-1)\rho^{m-2}, \tag{4-76}$$

which is not the case. Hence, $\eta''(t_0) \neq 0$ and, since it is convex, $\eta''(t_0) > 0$. Completely analogously, by substituting '=' by '>' in (4-75) and (4-76), one obtains that this is equivalent to $\phi_A(\rho) > m(m-1)\rho^{m-2}$.

□

We have made all preparations in order to prove the main result of this section. Recall once again that if the m–monic PFP P has a right PFF C, i. e.,

$$P(\lambda) = \left(\sum_{j=0}^{l-m} \lambda^j B_j\right)\left(\lambda^m I_n - \sum_{j=0}^{m} \lambda^j C_j\right) \quad \text{with} \quad C_j \geqslant 0 \ (j \in \langle m \rangle_0),$$

then by Lemma 4.9, the coefficient B_0 is a Z-matrix.

Theorem 4.28. *Let P be an m–monic PFP with*

$$P(\lambda) = \lambda^m I_n - A(\lambda) = \lambda^m I_n - \sum_{j=0}^{l} \lambda^j A_j$$

of degree l with irreducible A and let d be the index of phase imprimitivity of the graph $G_m(A_0, \ldots, A_l)$. Then exactly one of the following cases holds.

(i) $\phi_A(\tau) > \tau^m$ *for all* $\tau \geq 0$. *Then* P *has no Perron-Frobenius factor.*

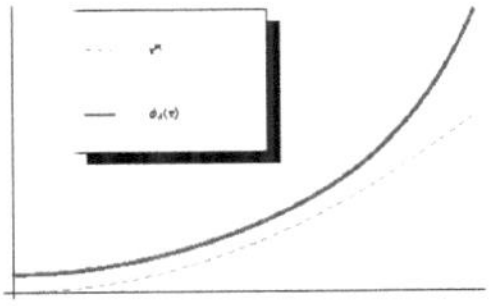

(ii) $\phi_A(\tau) > \tau^m$ *for all* $\tau > 0$ *and* $\phi_A(0) = 0$. *Then either* P *has no Perron-Frobenius factor or* B_0 *is not a regular* M*-matrix.*

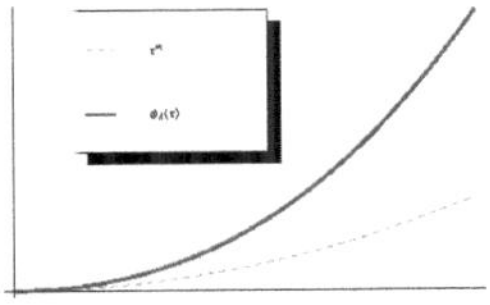

(iii) There exists exactly one $\rho > 0$ *with* $\phi_A(\rho) = \rho^m$. *If*

(a) $\phi'_A(\rho) < m\rho^{m-1}$, *then* P *has a spectral right Perron-Frobenius factor, i. e.,*

$$P(\lambda) = \widetilde{B}(\lambda)B_0C(\lambda)$$

with $\operatorname{spr}(C) = \rho$ *and* $\operatorname{spr}(\operatorname{rev}\widetilde{B}) = 0$.

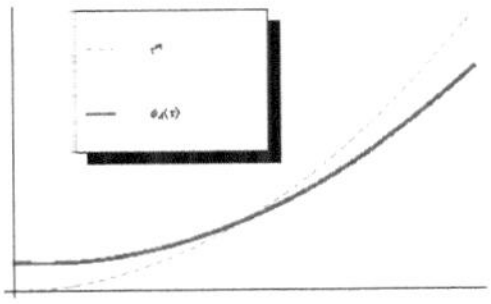

In particular, P *has*

- $mn - d$ *eigenvalues in* $\mathbb{D}_\rho$ *(counting multiplicities),*
- d *algebraically simple eigenvalues on* $\mathbb{T}_\rho$ *at the dth roots of* ρ^d *and*
- $(l - m)n$ *eigenvalues at* ∞ *(counting multiplicities).*

(b) $\phi'_A(\rho) > m\rho^{m-1}$, *then* P *has a spectral right Perron-Frobenius factor, i. e.,*

$$P(\lambda) = \widetilde{B}(\lambda)B_0C(\lambda)$$

with $\operatorname{spr}(C) = 0$ *and* $\operatorname{spr}(\operatorname{rev}\widetilde{B}) = \frac{1}{\rho}$.

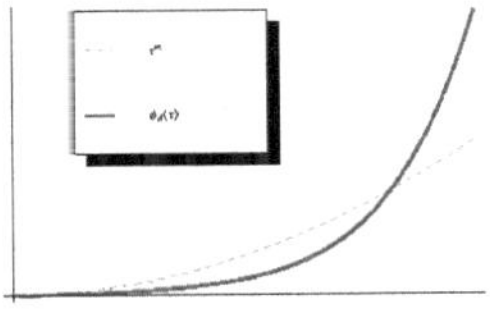

In particular, P *has*

- 0 *as an eigenvalue with multiplicity* mn,
- d *algebraically simple eigenvalues on* $\mathbb{T}_\rho$ *at the dth roots of* ρ^d *and*
- $(l-m)n-d$ *eigenvalues outside* $\overline{\mathbb{D}}_r$ *(including* ∞*, counting multiplicities).*

(c) $\phi'_A(\rho) = m\rho^{m-1}$, *then* P *has a right Perron-Frobenius factor, i. e.,*

$$P(\lambda) = \widetilde{B}(\lambda)B_0C(\lambda)$$

with

- $\operatorname{spr}(C) = \rho$, $\operatorname{spr}(\operatorname{rev}\widetilde{B}) = \frac{1}{\rho}$,

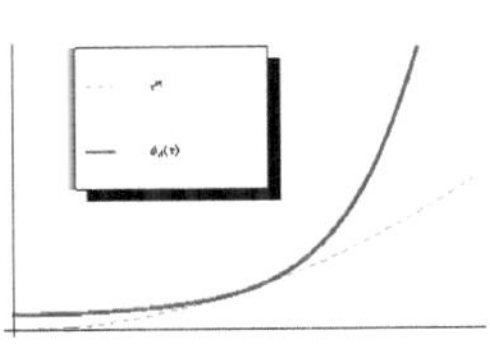

- B_0 *is a nonsingular* M*-matrix and*
- $\phi''_A(\rho) > m(m-1)\rho^{m-2}$.

Furthermore, P *has*

- $mn - d$ *eigenvalues in* $\mathbb{D}_\rho$ *(counting multiplicities),*
- d *eigenvalues of geometric multiplicity 1 and algebraic multiplicity 2 on* $\mathbb{T}_\rho$ *at the dth roots of* ρ^d *and*
- $(l-m)n - d$ *eigenvalues outside* $\overline{\mathbb{D}}_\rho$*(including* ∞*, counting multiplicities).*

(*iv*) *There exist exactly two numbers* $\rho_2 > \rho_1 > 0$ *with* $\phi_A(\rho_j) = \rho_j^m$ *for* $j = 1, 2$. *Then* P *has a spectral right Perron-Frobenius factor*

$$P(\lambda) = \widetilde{B}(\lambda)B_0C(\lambda)$$

with respect to any $\tau \in (\rho_1, \rho_2)$ *with* $\operatorname{spr}(C) = \rho_1$, $\operatorname{spr}(\operatorname{rev} B) = \frac{1}{\rho_2}$ *and* B_0 *is a nonsingular* M*-matrix.* C *is independent of* τ.

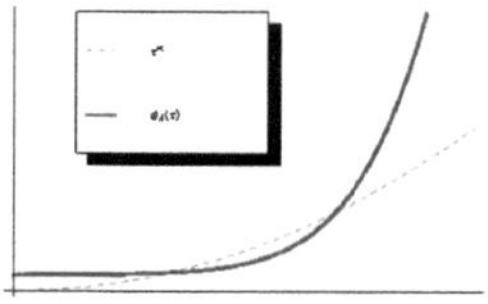

In particular, P *has*

- $mn - d$ *eigenvalues in* $\mathbb{D}_{\rho_1}$ *(counting multiplicities)*
- d *simple eigenvalues on* $\mathbb{T}_{\rho_1}$ *and* d *simple eigenvalues on* $\mathbb{T}_{\rho_2}$ *at the dth roots of* ρ_1^d *and* ρ_2^d*, respectively and*
- $mn-d$ *eigenvalues outside* $\overline{\mathbb{D}}_{\rho_2}$ *(including* ∞ *and counting multiplicities).*

(*v*) $\phi_A(\tau) < \tau^m$ *for all* $\tau > 0$. *Then* P *has a spectral right Perron-Frobenius factor, i. e.,*

$$P(\lambda) = \widetilde{B}(\lambda)B_0C(\lambda)$$

with $\operatorname{spr}(C) = 0$, $\operatorname{spr}(\operatorname{rev} \widetilde{B}) = 0$ *and* B_0 *is a nonsingular* M*-matrix. In particular,* P *has*

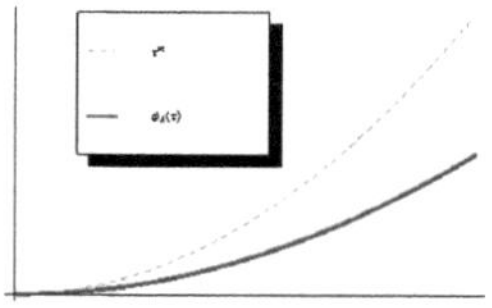

- 0 *as eigenvalue with multiplicity* mn *and*
- ∞ *as eigenvalue with multiplicity* $(l-m)n$.

(*vi*) $\phi_A(\tau) = \tau^m$ *for all* $\tau \geqslant 0$. *Then* P *has a right Perron-Frobenius factor and the left factor* B *is not invertible for all* $\lambda \in \mathbb{C}$*, hence,* P *is singular.*

Proof. (*i*) Suppose that P has a right Perron-Frobenius factor $C(\lambda)$. From $\phi_A(\tau) > \tau^m$ for all $\tau \geqslant 0$ and Proposition 4.25 (*i*) it follows that $\operatorname{spr} C = 0$. Hence, by Corollary 3.6 the corresponding right root $\mathcal{C}$ of the canonical reduction $\mathcal{P}$ of P to degree $l-m+1$ according to Corollary 3.3 is nilpotent. By [Rau92, Proposition 2.1], $\operatorname{spr}\left(\sum_{j=0}^{m-1} \tau^j C_j\right) = 0$ for all $\tau > 0$. Hence, $0 = \operatorname{spr}\left(\sum_{j=0}^{m-1} C_j\right) \geqslant \operatorname{spr} C_0$, i. e., $\operatorname{spr} C_0 = 0$. Since $\mathcal{C}$ satisfies (4-50) we have $\mathcal{A}_0 \leqslant \mathcal{C}$ and, therefore, by the structure of A_0 and C, $A_0 \leqslant C_0$. Since A_0 and C_0 are entrywise nonnegative we have $\operatorname{spr} A_0 \leqslant \operatorname{spr} C_0 = 0$, i. e., $\phi_A(0) = \operatorname{spr} A_0 = 0$ in contradiction to $\phi_A(\tau) > \tau^m$ for all $\tau \geqslant 0$. So P has no PFF.

(*ii*) Suppose that P has a right PFF, i. e., in particular

$$P(\lambda) = \widetilde{B}(\lambda) B_0 C(\lambda)$$

with C monic and $\widetilde{B}$ comonic. Suppose furthermore that B_0 is a nonsingular $M-$matrix.

By Proposition 4.25, $\operatorname{spr}(C) = 0$, thus $C(\tau)$ is invertible for all $\tau > 0$.

We have

$$P(\lambda) = \lambda^{l-m} \operatorname{rev} \widetilde{B}(1/\lambda) B_0 C(\lambda),$$

where $\operatorname{rev} \widetilde{B}(\lambda) = \lambda^{l-m} I_n - \sum_{j=1}^{l-m} \lambda^{l-m-j} \widetilde{B}_j$.

Suppose that $\rho = \operatorname{spr}(\operatorname{rev} \widetilde{B}) > 0$. Clearly, since $\operatorname{rev} \widetilde{B}$ is monic, $\rho < \infty$ and from Proposition 1.24 it follows that $\rho^{l-m} = \operatorname{spr}\left(\sum_{j=1}^{l-m} \rho^{l-m-j} \widetilde{B}_j\right)$. Hence, since $\sum_{j=1}^{l-m} \rho^{l-m-j} \widetilde{B}_j$ is a nonnegative matrix, there exists a nonnegative, nonzero vector v such that $\left(\sum_{j=1}^{l-m} \rho^{l-m-j} \widetilde{B}_j\right)^T v = \rho^{l-m} v$, or equivalently $\operatorname{rev} \widetilde{B}^T(\rho) v = 0$ Therefore, we have

$$P^T(1/\rho) v = \frac{1}{\rho^{l-m}} C^T(1/\rho) B_0^T \operatorname{rev} \widetilde{B}^T(\rho) v = 0,$$

i. e., v is an eigenvector for P^T to the eigenvalue $\frac{1}{\rho}$. Then, due to Proposition 4.18,

$$\left(\frac{1}{\rho}\right)^m = \operatorname{spr} A^T(1/\rho) = \operatorname{spr} A(1/\rho) = \phi_A(1/\rho),$$

which is a contradiction to the assumption. Hence, $\operatorname{spr}(\operatorname{rev} \widetilde{B}) = 0$, $\widetilde{B}$ has no eigenvalues in $\mathbb{C}$ and P is invertible for all $\tau > 0$. $P(\tau)$ is a Z-matrix and $\operatorname{spr} A(\tau) = \phi_A(\tau) > \tau^m$, so for all $\tau > 0$ the inverse $P(\tau)^{-1}$ is not nonnegative.

On the other hand

$$P(\tau)^{-1} = \frac{1}{\tau^{l-m}} C(\tau)^{-1} B_0^{-1} \operatorname{rev} B(1/\tau)^{-1},$$

where by Proposition 1.26 $C(\tau)^{-1} \geqslant 0$ and $\operatorname{rev} B(1/\tau)^{-1} \geqslant 0$. Hence, $P(\tau)^{-1}$ is nonnegative. This is a contradiction.

(*iii*) (a) From Proposition 4.25 (*ii*) it follows that P has a right PFF. Pick the minimal one and denote it by C, such that $P(\lambda) = B(\lambda) C(\lambda)$.

By assumption, for $\tau > \rho$ we have $\phi_A(\tau) < \tau^m$, since $\phi_A'(\rho) < m\rho^{m-1}$ and ρ is the only positive number such that $\phi_A(\rho) = \rho^m$ and ϕ_A is continuous. This implies that $P(\tau)$ is a nonsingular M-matrix for all $\tau > \rho$ and, therefore, $B(\tau)$ is invertible for all $\tau > \rho$.

The assumption and Proposition 1.15 (*ii*) imply that P has exactly mn eigenvalues in $\overline{\mathbb{D}}_\rho$ and, therefore, has exactly mn finite eigenvalues. C has exactly mn eigenvalues all of which are finite, since C is of degree m and it is monic. So all of the mn finite eigenvalues of P are eigenvalues of C

and furthermore $\operatorname{spr}(C)=\rho$, since ρ is the largest finite eigenvalue of P in absolute value.

This also implies that $\operatorname{spr}(\operatorname{rev} B) < \frac{1}{\rho}$, i. e., $B(\tau)$ is invertible for all $\tau \in [0,\rho]$.

Hence, $B(\tau)$ is invertible for all $\tau \geqslant 0$ which is equivalent to $\operatorname{spr}(\operatorname{rev} B) = 0$. Note that this also implies that B_0 is nonsingular.

Consider the factorization $\mathcal{P}(\lambda) = \left(\sum_{j=0}^{l-m} \lambda^j \mathcal{B}_j\right)(\lambda I_{mn} - \mathcal{C})$ of the canonical reduction $\mathcal{P}$ corresponding to the factorization $P(\lambda) = B(\lambda)C(\lambda)$ of P. Note that then $\mathcal{B}_0 = \begin{bmatrix} B_0 & * & * & * \\ & I_n & & \\ & & \ddots & \\ & & & I_n \end{bmatrix}$, see (3-43), so $\mathcal{B}_0$ is nonsingular.

From Corollary 4.3 it follows that $\mathcal{B}_0$ is a Z-matrix and so is B_0. From (4-53) we know that

$$\mathcal{B}_0 = I_{mn} - \sum_{j=0}^{l-1} \mathcal{A}_{j+1}\mathcal{C}^j.$$

$A(\rho)$ is irreducible, so let $v \in \mathbb{C}^n$, $v > 0$ be such that $P(\rho)v = 0$ and set $\hat{v} = \Lambda(\rho)v > 0$, where Λ is the operator from (3-45). Then $\mathcal{A}(\rho)\hat{v} = \rho\hat{v}$ and by Proposition 4.5 $\mathcal{C}\hat{v} \leqslant \rho\hat{v}$. Hence,

$$\sum_{j=0}^{l-1} \mathcal{A}_{j+1}\mathcal{C}^j\hat{v} \leqslant \frac{1}{\rho}\sum_{j=0}^{l-1} \mathcal{A}_{j+1}\rho^{j+1}\hat{v} \leqslant \frac{1}{\rho}\sum_{j=0}^{l} \mathcal{A}_j\rho^j\hat{v} = \frac{1}{\rho}\mathcal{A}(\rho)\hat{v} = \hat{v},$$

so $\operatorname{spr}\left(\sum_{j=0}^{l-1} \mathcal{A}_{j+1}\mathcal{C}^j\right) \leqslant 1$. Suppose that equality holds here. Then we can find a $\hat{w} \in \mathbb{C}^{mn}$, $\hat{w} \neq 0$ such that $\sum_{j=0}^{l-1} \mathcal{A}_{j+1}\mathcal{C}^j\hat{w} = \hat{w}$, which implies that $\mathcal{B}_0\hat{w} = 0$, thus $\hat{w} = 0$, since $\mathcal{B}_0$ is nonsingular. Therefore, $\operatorname{spr}\left(\sum_{j=0}^{l-1} \mathcal{A}_{j+1}\mathcal{C}^j\right) < 1$ and hence, $\mathcal{B}_0$ and B_0 are nonsingular M-matrices.

By Proposition 1.15(ii) it follows that P has exactly mn eigenvalues in the open disc $\overline{\mathbb{D}}_\rho$, d of which lie on the circle $\mathbb{T}_\rho$ at the d-th roots of ρ^d due to Theorem 4.23. The remaining $(l-m)n$ eigenvalues of P must belong to B, which has no eigenvalues in $\mathbb{C}$ due to $\operatorname{spr}(\operatorname{rev} B) = 0$. Hence, these are eigenvalues at infinity. Since the matrix polynomials B and $\widetilde{B} = BB_0^{-1}$ have the same eigenvalues, the assertion is proved.

(b) By assumption we have that $\phi_A(\tau) < \tau^m$ for all $\tau \in (0,\rho)$, since ρ is the only positive number with $\phi_A(\rho) = \rho^m$ and ϕ_A is continuous. Let $\mathcal{P}(\lambda) = \lambda I_{mn} - \mathcal{A}(\lambda)$ be the canonical reduction of P. Then $\phi_{\mathcal{A}}(\tau) < \tau$ for all $\tau \in (0,\rho)$. Then by Proposition 4.7 for each $\tau \in (0,\rho)$, $\mathcal{P}$ has a spectral right root and hence, P has a spectral right PFF C_τ w.r.t. $\mathbb{T}_\tau$. $\phi_A(\tau) < \tau^m$ then implies that $\operatorname{spr}(C_\tau) = 0$, i. e., $\sigma(C_\tau) = \{0\}$ and the eigenvalue zero has the algebraic multiplicity mn. By [Mar88, Lemma 22.8], $C = C_\tau$ does not depend on τ.

ρ is an eigenvalue of P, since $\phi_A(\rho) = \rho^m$. Hence, $\operatorname{spr}(\operatorname{rev} \widetilde{B}) = \frac{1}{\rho}$. By Theorem 4.23 the number of eigenvalues of P on $\mathbb{T}_\rho$ is d and they lie at the dth roots of ρ^d. The last statement on the eigenvalues follows immediately.

(c) The statements on the second derivative of ϕ_A and on the eigenvalues of P on $\mathbb{T}_\rho$ directly follow from Theorem 4.23 and Proposition 4.27.

(*iv*) From Theorem 4.23 it follows that P has eigenvalues on $\mathbb{T}_{\rho_1}$ and $\mathbb{T}_{\rho_2}$ at the dth roots of ρ_1^d and ρ_2^d, respectively. For any $\tau \in (\rho_1, \rho_2)$ we have $\phi_A(\tau) < \tau^m$, so by Theorem 4.1, P has a spectral PFF w.r.t. $\mathbb{T}_\tau$ i. e. $P(\lambda) = \widetilde{B}(\lambda)B_0C(\lambda)$, with $\sigma(C) \subset \mathbb{D}_\tau$ and $\sigma(\widetilde{B}) \subset \mathbb{C} \setminus \overline{\mathbb{D}}_\tau$. Since there are no eigenvalues of P in the annulus $\mathbb{A}_{\rho_1,\rho_2}$, the eigenvalues on $\mathbb{T}_{\rho_1}$ are peripheral eigenvalues of C and the eigenvalues on $\mathbb{T}_{\rho_2}$ are peripheral eigenvalues of rev $\widetilde{B}$. Proposition 4.25 (*iii*) and (*iv*) imply that these eigenvalues are algebraically simple. The remaining part of the assertion follows immediately and from the uniqueness of the spectral PFF with a given spectrum (see [Mar88, Lemma 22.8]).

(*v*) By Theorem 4.1, P has a spectral PFF C_τ w.r.t. to $\mathbb{T}_\tau$ for any $\tau > 0$ and Proposition 4.25 (*i*) implies that $\operatorname{spr}(C_\tau) = 0$. Again, $C = C_\tau$ is independent of τ and so $\operatorname{spr}(C) = 0$ and also the left factor $\widetilde{B}$ does not depend on τ, since it is uniquely determined by C. So from $\sigma(\widetilde{B}) \subset \mathbb{C} \setminus \overline{\mathbb{D}}_\tau$ for all $\tau > 0$ it follows that $\widetilde{B}$ has no finite eigenvalues, hence, all its $(l-m)n$ eigenvalues are at infinity.

(*vi*) From Proposition 4.25 it follows that P has a right PFF C, $P(\lambda) = B(\lambda)C(\lambda)$. Since each $\tau > 0$ is an eigenvalue of P, $\sigma(P) = \mathbb{C}$, i. e., P is singular. C is monic, so C has finitely many eigenvalues and B has to be singular.

□

Note that the PFF in case (*iii*)(*c*) of Theorem 4.28 is not a spectral PFF in the sense of our definition of the beginning of this chapter.

We now give a few examples concerning Theorem 4.28 and start with the special case of scalar polynomials.

Example 4.29. Consider the scalar polynomial $p(\lambda) = \lambda^m - a(\lambda)$ with $a(\lambda) = \sum_{j=0}^{l} \lambda^j a_j$, $l > m$, $a_j \geqslant 0$ and $\sum_{j=0}^{l} a_j > 0$. Let $d = \gcd\{m - j : a_j > 0\}$.
p is an irreducible 1×1 m–monic polynomial and d is the index of phase imprimitivity of the graph $G_m(a_0, \ldots, a_l)$, which has only one vertex. We apply Theorem 4.28 to study the roots of p and the existence of factorizations

$$p(\lambda) = \left(b_0 - \sum_{j=1}^{l-m} \lambda^j b_j\right)\left(\lambda^m - \sum_{j=0}^{m-1} \lambda^j c_j\right), \tag{4-77}$$

such that $b_1, b_2, \ldots, b_{l-m}, a_0, a_1, \ldots, a_m \geqslant 0$ and such that the roots of $b(\lambda) = b_0 - \sum_{j=1}^{l-m} \lambda^j b_j$ and $c(\lambda) = \lambda^m - \sum_{j=0}^{m-1} \lambda^j c_j$ are separated by a certain circle. Note that the function ϕ_a in the scalar case can be written as $\phi_a(\tau) = a(\tau)$, $\tau > 0$. This also implies that the function $a : (0, \infty) \to (0, \infty)$ is geometrically convex. From Theorem 4.28 it follows that the following five cases can occur.

(*i*) $a(\tau) > \tau^m$ for all $\tau \geq 0$. Then p has no factorization as in (4-77)

(*ii*) $a(\tau) > \tau^m$ for all $\tau > 0$ and $a(0) = 0$. Then either p has no factorization as in (4-77) or $b_0 \leqslant 0$.

(*iii*) There exists exactly one $\rho > 0$ with $a(\rho) = \rho^m$. If

  (a) $a'(\rho) > m\rho^{m-1}$, then p has a factorization as in (4-77), such that c only has 0 as a root with order m, b has d simple roots on $\mathbb{T}_\rho$, which are precisely the dth roots of ρ^d and $l-m-d$ roots outside of $\overline{\mathbb{D}}_\rho$ (counting orders).

  (b) $a'(\rho) = m\rho^{m-1}$, then p has a factorization as in (4-77) such that c has $m-d$ roots in $\mathbb{D}_\rho$, b has $l-m-d$ roots outside $\overline{\mathbb{D}}_\rho$ and both, b and c have simple roots on $\mathbb{T}_\rho$, respectively, which coincide with the dth roots of ρ^d. Furthermore, we have $a''(\rho) > m(m-1)\rho^{m-2}$.

(*iv*) There exist exactly two numbers $\rho_2 > \rho_1 > 0$ with $a(\rho_j) = \rho_j^m$ for $j = 1, 2$. Then p has a factorization as in (4-77), c has $m-d$ roots in $\mathbb{D}_\rho$ (counting orders) and d simple roots on $\mathbb{T}_{\rho_1}$ which coincide with the dth roots of ρ_1^d. b has $l-m-d$ roots outside $\overline{\mathbb{D}}_\rho$ and d simple roots on $\mathbb{T}_{\rho_1}$ which coincide with the dth roots of ρ_2^d.

The next two examples can be found in [FN05a, Example 4.11]. They show that in case (*ii*) of Theorem 4.28 both possibilities - either P has no PFF or B_0 is not a regular M-matrix - can occur.

Example 4.30. Let $n \in \mathbb{N}$ and let $A(\lambda) = \lambda^2 A_2 + \lambda I_n$, where $A_2 \in \mathbb{R}^{n,n}$ is strictly positive and irreducible. Then we have

$$P(\lambda) = \lambda I_n - A(\lambda) = -\lambda^2 A_2 = (0 - \lambda A_2)(\lambda I_n - 0),$$

$\phi_A(\tau) = \tau + \tau^2 \operatorname{spr}(A_2) > \tau$ for $\tau > 0$, and P has the minimal right PFF $\lambda I_n - 0$, but $B_0 = 0$ is not invertible.

Example 4.31. Consider again Example 1.5, i. e.,

$$A(\lambda) = \begin{bmatrix} \lambda^2 & 1 \\ \lambda p & \lambda^2 \end{bmatrix},$$

with $p > 0$. Then, with the notation from Example 1.5, $\phi_A(\tau) = \operatorname{spr} A(\tau) > \tau$ for all $\tau > 0$ is equivalent to $\xi - \xi^3 < \sqrt{p}$ for all $\xi > 0$, which holds if and only if $\xi_0 - \xi_0^3 < \sqrt{p}$, or $p > \frac{4}{27}$.

We will show that for $p > \frac{4}{27}$, $P(\lambda) = \lambda I_2 - A(\lambda)$ has no PFP. Suppose that C is a nonnegative root of P, i. e. $P(\lambda) = (\lambda B_1 + B_0)(\lambda I_2 - C)$. Then C is nonzero, since $A_0 = \left[\begin{smallmatrix} 0 & 1 \\ 0 & 0 \end{smallmatrix}\right]$ is nonzero and by Proposition 4.25 (*i*), C is nilpotent. Therefore, $C = \left[\begin{smallmatrix} 0 & 0 \\ c & 0 \end{smallmatrix}\right]$ or $C = \left[\begin{smallmatrix} 0 & c \\ 0 & 0 \end{smallmatrix}\right]$ with $c > 0$.

From the first case it follows that $\left[\begin{smallmatrix} 0 & 1 \\ 0 & 0 \end{smallmatrix}\right] = B_0 C = \left[\begin{smallmatrix} \alpha & 0 \\ \beta & 0 \end{smallmatrix}\right]$, with some $\alpha, \beta \in \mathbb{R}$, which is a contradiction.

If we suppose the second case and set $B_0 = \left[\begin{smallmatrix} \alpha_0 & \beta_0 \\ \gamma_0 & \delta_0 \end{smallmatrix}\right]$, $B_1 = \left[\begin{smallmatrix} \alpha_1 & \beta_1 \\ \gamma_1 & \delta_1 \end{smallmatrix}\right]$, then we have $\left[\begin{smallmatrix} 0 & 1 \\ 0 & 0 \end{smallmatrix}\right] = B_0 C = \left[\begin{smallmatrix} 0 & \alpha_0 c \\ 0 & \gamma_0 c \end{smallmatrix}\right]$, hence, $B_0 = \left[\begin{smallmatrix} c^{-1} & \beta_0 \\ 0 & \delta_0 \end{smallmatrix}\right]$ and therefore

$$\left[\begin{smallmatrix} 1 & 0 \\ -p & 1 \end{smallmatrix}\right] = I_2 - A_1 = B_0 - B_1 C = \left[\begin{smallmatrix} c^{-1} & \beta_0 - \alpha_1 c \\ 0 & \delta_0 - \gamma_1 c \end{smallmatrix}\right],$$

which, again, is a contradiction. Hence, P has no PFF.

4.4 The operator equation $X = \sum_{j=0}^{l} A_j X^j$

In this section we show that the fixpoint iteration from Section 4.1 does also converge in a more general setting under slightly stronger conditions. These conditions appear in [Mar88], where they imply the existence of a spectral factorization of an operator polynomial.

Proposition 4.32. *Let $\mathfrak{A}$ be a complex Banach algebra and let $P(\lambda) = \lambda I - \sum_{j=0}^{l} \lambda^j A_j$ be a 1-monic algebra polynomial with $A_j \in \mathfrak{A}$. Suppose that there exists a $\rho > 0$ such that $\tilde{\phi}_A(\rho) := \sum_{j=0}^{l} \rho^j \|A_j\| < \rho$.*
Then the fixpoint iteration $(X_k)_{k\in\mathbb{N}} \subset \mathfrak{A}$ defined via

$$X_0 = 0, \quad X_{k+1} = \sum_{j=0}^{l} A_j X_k^j \tag{4-78}$$

converges.

Proof. W.l.o.g. we may assume that $\gamma := \tilde{\phi}'_A(\rho) = \sum_{j=1}^{l} j\rho^{j-1}\|A_j\| < 1$. Otherwise, there exists a nonnegative $\rho_0 < \rho$ such that $\tilde{\phi}_A(\rho_0) = \rho_0$ and $\tilde{\phi}'_A(\rho_0) < 1$. We can choose a $\rho_1 > \rho_0$ near ρ_0 such that $\tilde{\phi}_A(\rho_1) < \rho_1$ and still $\tilde{\phi}'_A(\rho_1) < 1$.

We will show that $(X_k)_{k\in\mathbb{N}}$ is a Cauchy sequence.

We have that $\|X_k\| < \rho$ for all $k \in \mathbb{N}$. Indeed, clearly, $\|X_0\| < \rho$. Suppose that for some $k \in \mathbb{N}$, $\|X_k\| < \rho$. Then $\|X_{k+1}\| \leqslant \sum_{j=0}^{l} \|A_j\| \|X_k\|^j \leqslant \sum_{j=0}^{l} \|A_j\| \rho^j < \rho$, by assumption. For all $k \in \mathbb{N} \setminus \{0\}$ we have

$$\begin{aligned}
\|X_k^j - X_{k-1}^j\| &= \left\| \sum_{\nu=0}^{j-1} X_k^{j-\nu} X_{k-1}^{\nu} - X_k^{j-1-\nu} X_{k-1}^{\nu+1} \right\| \\
&= \left\| \sum_{\nu=0}^{j-1} X_k^{j-1-\nu} (X_k - X_{k-1}) X_{k-1}^{\nu} \right\| \\
&\leqslant \sum_{\nu=0}^{j-1} \|X_k^{j-1-\nu}\| \|X_k - X_{k-1}\| \|X_{k-1}^{\nu}\| \\
&< \sum_{\nu=0}^{j-1} \rho^{j-1} \|X_k - X_{k-1}\| = j\rho^{j-1} \|X_k - X_{k-1}\|.
\end{aligned}$$

Hence, from the definition of the iteration it follows that for all $k \in \mathbb{N}$

$$\|X_{k+1} - X_k\| \leqslant \sum_{j=0}^{l} j\rho^{j-1} \|A_j\| \|X_k - X_{k-1}\| = \gamma \|X_k - X_{k-1}\|.$$

Let $p \in \mathbb{N}$ be such that $X_p \neq 0$ and $X_k = 0$ for $k < p$. Then we have for all $k \geqslant p$

$$\|X_{k+1} - X_k\| \leqslant \gamma^{k-p+1} \|X_p - X_{p-1}\| = \gamma^{k-p+1} \|X_p\|.$$

□

Notice that the condition $\phi_A(\rho) < \rho$ from Theorem 4.8 is weaker than the condition $\tilde{\phi}_A(\rho) = \sum_{j=0}^{l} \rho^j \|A_j\| < \rho$ from Proposition 4.32. Indeed, from the latter it follows that

$$\phi_A(\rho) = \sup_{|\lambda|=\rho} \operatorname{spr} A(\lambda) \leqslant \sup_{|\lambda|=\rho} \|A(\lambda)\| \leqslant \sup_{|\lambda|=\rho} \sum_{j=0}^{l} \rho^j \|A_j\| = \tilde{\phi}_A(\rho) < \rho.$$

Let $X \in \mathfrak{A}$ be the limit of the fixpoint iteration (4-78). From [Mar88, Lemma 22.9] it then follows that under the conditions of Proposition 4.32, $\lambda I - X$ is a right divisor of the polynomial P. A. S. Markus obtained this result in [Mar88, Section 23] as a consequence of an abstract factorization result in decomposing Banach algebras, see [GKS03, Theorem 2.5]. Proposition 4.32 implies the existence of a divisor of P via basic operator theoretical concepts.

5 Computing spectral factorizations of m–monic matrix polynomials with a cyclic reduction algorithm

Suppose that for the m–monic Perron-Frobenius polynomial (PFP, see Chapter 4)

$$P(\lambda) = \lambda^m I_n - A(\lambda) = \lambda^m I_n - \sum_{j=0}^{l} \lambda^j A_j$$

the conditions of Theorem 4.1 hold. In particular, there exists a $\rho > 0$ such that $\operatorname{spr}(A(\rho)) < \rho^m$. In this section, by writing ρ we always refer to this number. Let

$$P(\lambda) = B(\lambda) B_0 C(\lambda) = \left(I_n - \sum_{j=1}^{l-m} \lambda^j B_j \right) B_0 \left(\lambda^m I_n - \sum_{j=0}^{m-1} \lambda^j C_j \right) \tag{5-79}$$

be the spectral factorization according to Theorem 4.1. It is, furthermore, assumed that A is irreducible, i.e. that $A(\tau)$ is an irreducible matrix for all $\tau > 0$, see Section 4.3.

In this section we describe how a cyclic reduction method for certain Markov process type problems can be used to explicitly calculate the factors B, B_0 and C. Cyclic reduction (CR) was designed as a powerful direct method for solving certain structured matrix problems, in particular block tridiagonal block Toeplitz systems encountered in the finite differences discretization of the Poisson equation over a rectangle. Several modifications have been developed, for instance for solving linear systems arising from the discretization of boundary value ODEs. The cyclic reduction method has been rediscovered as a quadratically convergent iterative algorithm for solving certain infinite systems and nonlinear matrix equations associated with stochastic processes, see e.g. [AP97], [BLM05], [BM09], [BG94], [BGN70], [Hel76], [Swe88].

We will refer to the CR method presented in [BLM05, Chapter 7.4]. In this book, the CR method is used to compute the minimal nonnegative solution $G_{\min}$ of the matrix equation

$$X = \sum_{i=-1}^{\infty} A_i X^{i+1}, \tag{5-80}$$

where minimal means that if Y is any other solution of this equation, then $G_{\min} \leqslant Y$.

We will start with sketching the basic idea of CR (for a more detailed description of the method, see [BLM05]). Before we do that, we will briefly recall the notion of Toeplitz operators.

Associate with the matrix Laurent power series defined by $S(\lambda) = \sum_{j\in\mathbb{Z}} \lambda^j S_j$ the semi infinite block matrix

$$T_\infty[S] = \begin{bmatrix} S_0 & S_1 & S_2 & S_3 & \cdots \\ S_{-1} & S_0 & S_1 & S_2 & \ddots \\ S_{-2} & -S_{-1} & S_0 & S_1 & \ddots \\ S_{-3} & S_{-2} & S_{-1} & S_0 & \ddots \\ \vdots & \ddots & \ddots & \ddots & \ddots \end{bmatrix}. \tag{5-81}$$

This matrix is called a **block Toeplitz matrix**, since $(T_\infty[S])_{ij} = S_{j-i}$ for all $i,j \in \mathbb{N}$, where $(T_\infty[S])_{ij}$ denotes the $n \times n$ block in the i-th block row and j-th block column of $T_\infty[S]$.

An infinite block Toeplitz matrix

$$T = \begin{bmatrix} F_0 & F_{-1} & F_{-2} & \cdots \\ F_1 & F_0 & F_{-1} & \ddots \\ F_2 & F_1 & F_0 & \ddots \\ \vdots & \ddots & \ddots & \ddots \end{bmatrix}$$

with $F_j \in \mathbb{C}^{n,n}$ for all $j \in \mathbb{Z}$, induces a bounded operator on ℓ_n^2, where ℓ_n^2 denotes the space of all sequences $(x_k)_{k\in\mathbb{N}} \subset \mathbb{C}^n$ such that

$$\|x\| := \left(\sum_{k=1}^{\infty} \|x_k\|^2 \right)^{1/2} < \infty,$$

if and only if the entries of T are the Fourier coefficients of a matrix function with entries in $L^\infty(\mathbb{T})$, more precisely,

$$F_j = \frac{1}{2\pi} \int_0^{2\pi} F(e^{i\varphi}) e^{-ij\varphi} d\varphi,$$

for some matrix function $F : \mathbb{T} \to \mathbb{C}^{n,n}$ with entries in $L^\infty(\mathbb{T})$, see e.g. [BS99], [BLM05].

The basic idea of CR is to rewrite equation (5-80) as the semi-infinite linear system

$$\begin{bmatrix} I - A_0 & -A_1 & -A_2 & \cdots \\ -A_{-1} & I - A_0 & -A_1 & \ddots \\ & -A_{-1} & I - A_0 & \ddots \\ & & \ddots & \ddots \end{bmatrix} \begin{bmatrix} X \\ X^2 \\ X^3 \\ \vdots \end{bmatrix} = \begin{bmatrix} A_{-1} \\ 0 \\ \vdots \\ \vdots \end{bmatrix}.$$

After applying a suitable (so called even-odd-) permutation, one obtains the system

$$\begin{bmatrix} I - U_{1,1}^{(1)} & -U_{1,2}^{(1)} \\ -U_{2,1}^{(1)} & I - U_{2,2}^{(1)} \end{bmatrix} \begin{bmatrix} \hat{X}_+ \\ \hat{X}_- \end{bmatrix} = \begin{bmatrix} 0 \\ \hat{B} \end{bmatrix},$$

where $U_{1,1}^{(1)}, U_{2,2}^{(1)}, U_{1,2}^{(1)}, U_{2,1}^{(1)}$ are some semi infinite Toeplitz matrices, $\hat{X}_+ = \begin{bmatrix} X^2 \\ X^4 \\ \vdots \end{bmatrix}$, $\hat{X}_- = \begin{bmatrix} X \\ X^3 \\ \vdots \end{bmatrix}$ and $\hat{B} = \begin{bmatrix} A_{-1} \\ 0 \\ \vdots \end{bmatrix}$. In particular, $I - U_{1,1}^{(1)}$ is an upper block triangular matrix with $I_n - A_0$ on the diagonal. Under suitable conditions $U_{1,2}^{(1)}, U_{2,1}^{(1)}, I - U_{2,2}^{(1)}$ and $(I - U_{1,1}^{(1)})^{-1}$ represent bounded operators. Performing one step of block Gaussian elimination, one obtains

$$\begin{bmatrix} I - U_{1,1}^{(1)} & -U_{1,2}^{(1)} \\ 0 & H^{(1)} \end{bmatrix} \begin{bmatrix} \hat{X}_+ \\ \hat{X}_- \end{bmatrix} = \begin{bmatrix} 0 \\ \hat{B} \end{bmatrix}, \tag{5-82}$$

where $H^{(1)} = I - U_{2,2}^{(1)} - U_{2,1}^{(1)}(I - U_{1,1}^{(1)})^{-1}U_{1,2}^{(1)}$ is the Schur complement of $I - U_{2,2}^{(1)}$. Analyzing $H^{(1)}$, one finds that $H^{(1)}$ has block Hessenberg form (i.e., it has only zero blocks below the first subdiagonal) and block Toeplitz form except for the first block row. Hence, the lower system of (5-82), $H^{(1)}\hat{X}_- = \hat{B}$, can be written as

$$\begin{bmatrix} I - \hat{A}_0^{(1)} & -\hat{A}_1^{(1)} & -\hat{A}_2^{(1)} & \cdots \\ -A_{-1}^{(1)} & I - A_0^{(1)} & -A_1^{(1)} & \ddots \\ & -A_{-1}^{(1)} & I - A_0^{(1)} & \ddots \\ 0 & & \ddots & \ddots \end{bmatrix} \begin{bmatrix} X \\ X^3 \\ X^5 \\ \vdots \end{bmatrix} = \begin{bmatrix} A_{-1} \\ 0 \\ \vdots \\ \vdots \end{bmatrix},$$

where the semi infinite block Toeplitz matrix on the left hand side can be proved to define a bounded operator on ℓ_n^2, see [BLM05, Section 7.4]. Recursively repeating this process of permuting and one step of Gaussian elimination leads to the system

$$\begin{bmatrix} I - \hat{A}_0^{(k)} & -\hat{A}_1^{(k)} & -\hat{A}_2^{(k)} & \cdots \\ -A_{-1}^{(k)} & I - A_0^{(k)} & -A_1^{(k)} & \ddots \\ & -A_{-1}^{(k)} & I - A_0^{(k)} & \ddots \\ 0 & & \ddots & \ddots \end{bmatrix} \begin{bmatrix} X \\ X^{2^k+1} \\ X^{2\cdot 2^k+1} \\ \vdots \end{bmatrix} = \begin{bmatrix} A_{-1} \\ 0 \\ \vdots \\ \vdots \end{bmatrix}$$

after the k-th step. The Schur complements $H^{(k)}$ (analogous to $H^{(1)}$) which appear in each step are well defined bounded operators, see [BLM05, Section 7.4]. Hence, the minimal solution $G_{\min}$ of equation (5-80) satisfies

$$G_{\min} = \left(I - \hat{A}_0^{(k)}\right)^{-1} \left(A_{-1} + \sum_{j=1}^{\infty} \hat{A}_j^{(k)} G_{\min}^{j\cdot 2^k+1}\right).$$

This method heavily relies on the invertibility of the upper block triangular matrix $I - U_{1,1}^{(k)}$ which appears in each step after the permutation. We suppose for the moment that this is satisfied and will return to this question later. It then follows that for each $k \in \mathbb{N}$ also $I - \hat{A}_0^{(k)}$ is invertible, hence, the given representation of $G_{\min}$ is well defined.

It can be proved that $\sum_{j=1}^{\infty} \hat{A}_j^{(k)} G_{\min}^{j\cdot 2^k+1}$ converges quadratically to zero for $k \to \infty$ and that $(I - \hat{A}_0^{(k)})^{-1}$ is quadratically convergent, see [BLM05, Section 7.4.1], hence,

$$G^{(k)} = (I - \hat{A}_0^{(k)})^{-1} A_{-1}$$

is a valid approximation for $G_{\min}$.

The CR algorithm presented in [BLM05] is suited for some special type of Markov chains. Markov chains are stochastic processes, i.e., families $\{X_j : j \in \mathbb{N}\}$ of random variables X_j, which have the so-called Markov property, i.e., that X_j only depends on X_{j-1}.

Associated with this type of Markov process is a so-called **generating function** of the form

$$\mathcal{P}(\lambda) = \lambda I_n - \sum_{j=-1}^{\infty} \lambda^{j+1} \mathcal{A}_j, \tag{5-83}$$

where the coefficients $\mathcal{A}_j$ are certain entrywise nonnegative matrices such that $\sum_{j=-1}^{\infty} \mathcal{A}_j$ is irreducible and stochastic, i. e. $\sum_{j=-1}^{\infty} \mathcal{A}_j \mathbf{1}_n = \mathbf{1}_n$. In [BLM05] is shown that CR can be applied for the **substochastic** case, i. e. $\sum_{j=-1}^{\infty} \mathcal{A}_j \mathbf{1}_n < \mathbf{1}_n$, as well.

5.1 Transformation to a Markov problem

In order to apply CR as depicted above to our PFP P, it has to be transformed in an appropriate way to a suitable generating function of some Markov chain of the type (5-83). For a reason that will become clear later in this section, it is useful to reduce the m–monic polynomial to a 1–monic polynomial via the canonical reduction from Chapter 3. Roughly spoken, this allows us to easily read off the coefficients C_j of the right Perron-Frobenius factor (PFF, see Chapter 4) C of P. Therefore, we suppose in this section that P is a 1–monic matrix polynomial. The factorization of the original polynomial then can easily be recovered as is described in Section 3.2.

If we have $P(\lambda) = \lambda^m I_n - A(\lambda)$ and $\mathcal{P}(\lambda) = \lambda I_{mn} - \mathcal{A}(\lambda)$ is the corresponding canonical reduction, then, due to Section 3.1,

$$\mathcal{A}(\lambda) = \begin{bmatrix} \sum_{j=m-1}^{l} \lambda^{j-m+1} A_j & A_{m-2} & \cdots & A_0 \\ I & & & \\ & \ddots & & \\ & & I & 0 \end{bmatrix}$$

and therefore, unfortunately, $\mathcal{A}$ is in general not irreducible if A is. More precisely, we have the following Proposition. For its proof recall the definition of the associated directed graph of a matrix as given in Section 4.3.

Proposition 5.1. *$\mathcal{A}$ is irreducible if and only if A_0 has no zero columns.*

Proof. Fix any $\tau > 0$ and consider the associated directed graph $G_{\mathcal{A}(\tau)} = (E, V)$ of $\mathcal{A}(\tau) \in \mathbb{C}^{mn,mn}$.

We have that

$$\mathcal{A}(\tau) = \begin{bmatrix} \sum_{j=m-1}^{l} \tau^{j-m+1} A_j & A_{m-2} & \cdots & A_0 \\ I & & & \\ & \ddots & & \\ & & I & 0 \end{bmatrix}.$$

For clarity, arrange the mn vertices of $G_{\mathcal{A}(\tau)}$ in a rectangle of m rows and n columns. Enumerate them from the left to the right and from top to bottom.

Consider $\tilde{\mathcal{A}} = \begin{bmatrix} 0 & \cdots & \cdots & 0 \\ I & \ddots & & \\ & & I & 0 \end{bmatrix}$, obtained by setting $A_0, \ldots, A_l = 0$ in $\mathcal{A}(\tau)$. Then, due to the positions of the identities in $\tilde{\mathcal{A}}$, the graph $G_{\tilde{\mathcal{A}}}$ looks as follows.

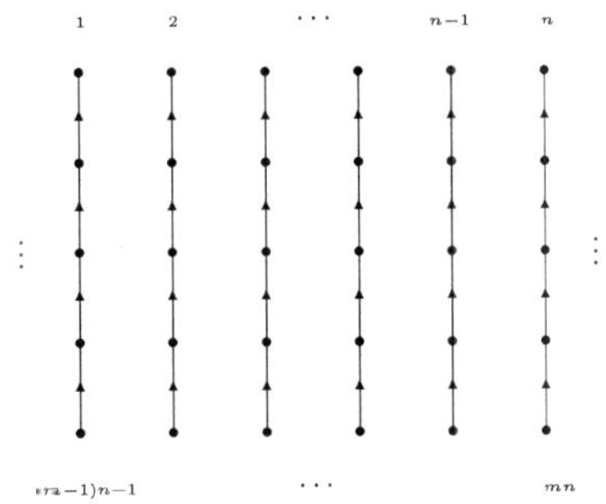

Figure 5: the graph $G_{\tilde{\mathcal{A}}}$

Therefore, for each $p, \tilde{p} \in \langle m-1 \rangle_0$, $\tilde{p} < p$ and for each $q \in \langle n \rangle$ there exists a path from $pn - q$ to $\tilde{p}n + q$.

The associated graph $G_{\mathcal{A}(\tau)}$ of $\mathcal{A}(\tau)$ looks the same with the difference that there are additional edges with starting vertices only in the first row, i.e. at the vertices $1, \ldots, n$. Note that $G_{\mathcal{A}(\tau)} = G_{\mathcal{A}(1)}$.

Obviously, if A_0 has a zero column, let us say column j, then there is no edge leading to vertex $(m-1)n + j$, and hence, the graph cannot be strongly connected. Thus $\mathcal{A}(\tau)$ is not irreducible.

Now suppose that A_0 has no zero columns. This implies that for each $j \in \{(m-1)n + 1, \ldots, mn\}$ there is an $i \in \langle n \rangle$ such that $(i, j) \in E$, in other words, each vertex in the last row has an incoming edge from at least one vertex of the first row.

Since by assumption $A(1) = \sum_{\nu=0}^{l} A_\nu$ is an irreducible matrix, we furthermore have that from each of the vertices in the first row there is a path to each column of the graph. More precisely, for each $i \in \langle n \rangle$ and each $j \in \langle n \rangle$ there is a path in $G_{\mathcal{A}(\tau)}$ from i to $pn + j$ for some $p \in \langle m-1 \rangle_0$. Indeed, fix any $i, j \in \langle n \rangle$ and let $i \to i_1 \to i_2 \to \cdots \to i_{k-1} \to j$ be a path in $A(1)$ from i to j. Then there are $p_1, p_2, \ldots, p_k \in \langle m-1 \rangle_0$ such that

$$i \to p_1 n + i_1 \to i_1 \to p_2 n + i_2 \to i_2 \to \cdots \to p_{k-1} n + i_{k-1} \to i_{k-1} \to p_k n + j$$

is a path in $G_{\mathcal{A}(\tau)}$.

Now, with these preparations we can give a path from any vertex $pn + q$ to any other vertex $rn + s$ ($p, r \in \langle m-1 \rangle_0$ and $q, s \in \langle n \rangle$). Let $\tilde{s} \in \langle n \rangle$ such that $(\tilde{s}, (m-1)n + s) \in E$, which exists since A_0 has no zero columns. Let furthermore

$\tilde{r} \in \langle m-1 \rangle_0$ be such that $q \to \cdots \to \tilde{r}n + \tilde{s}$ is a path in $G_{\mathcal{A}(\tau)}$. Then a path from $pn+q$ to $rn+s$ is

$$pn+q \to \cdots \to q \to \cdots \to \tilde{r}n+\tilde{s} \to \cdots \to \tilde{s} \to (m-1)n+s \to \cdots \to rn+s.$$

Hence, $\mathcal{A}(\tau)$ is irreducible. □

Clearly, in general, we cannot expect A_0 to have no zero columns and hence, $\mathcal{A}$ of the canonical reduction is in general not irreducible. But still an important property of the irreducible A is preserved as the following proposition states.

Proposition 5.2. *Let $P(\lambda) = \lambda^m I_n - A(\lambda)$ be an m–monic $n \times n$ PFP such that A is irreducible and let $\mathcal{P}(\lambda) = \lambda I_{mn} - \mathcal{A}(\lambda)$ be the canonical reduction of P. Then, for any $\tau > 0$, there exists a strictly positive eigenvector of $\mathcal{A}(\tau)$ corresponding to the eigenvalue* $\operatorname{spr}(\mathcal{A}(\tau))$.

Proof. Since $\mathcal{A}(\tau)$ is an entrywise nonnegative matrix, it has an entrywise nonnegative eigenvector $\hat{u} \in \mathbb{R}^{mn}$ corresponding to the eigenvalue $0 < r = \operatorname{spr}(\mathcal{A}(\tau))$. Write $\hat{u} = \begin{bmatrix} u_1 \\ \vdots \\ u_m \end{bmatrix}$ with $u_1, \ldots, u_m \in \mathbb{R}^n$. Hence, we have that

$$\begin{bmatrix} \sum\limits_{j=m-1}^{l} \tau^{j-m+1} A_j & A_{m-2} & \cdots & A_0 \\ I & & & \\ & \ddots & & \\ & & I & 0 \end{bmatrix} \begin{bmatrix} u_1 \\ \vdots \\ u_m \end{bmatrix} = r \begin{bmatrix} u_1 \\ \vdots \\ u_m \end{bmatrix}.$$

Solving this system, starting with the last row, we obtain $u_{m-j} = r^j u_m$ for $j \in \langle m-1 \rangle_0$. The first block row now reads

$$\tilde{A} u_m = \left(\sum_{j=m-1}^{l} r^{m-1} \tau^{j-m+1} A_j + r^{m-2} A_{m-2} + \cdots + r A_1 + A_0 \right) u_m = r^{m-1} u_m.$$

Now, the matrix $\tilde{A}$ on the left hand side of the equation is irreducible, since A is irreducible and therefore, u_m is strictly positive and r^{m-1} is the spectral radius of $\tilde{A}$, see e.g. [Min88, Theorem I.2.2, p.7] and [Min88, Theorem I.4.4, p.16]. But since $\hat{u} = \begin{bmatrix} r^{m-1} \\ \vdots \\ r \\ 1 \end{bmatrix} \otimes u_m$, $\hat{u}$ is strictly positive as well. □

As mentioned above, and in view of Proposition 5.2 we suppose from now on that P is a 1–monic PFP of degree l and that there exists a $\rho > 0$ such that $\operatorname{spr}(A(\rho)) < \rho$ and a strictly positive eigenvector corresponding to the eigenvalue $\operatorname{spr}(A(\rho))$ of $A(\rho)$.

The main step of the transformation to a Markov problem consists of a suitable scaling of $P(\lambda) = \lambda I_n - A(\lambda)$ into a polynomial $\bar{P}(\lambda) = \lambda I - \bar{A}(\lambda)$ such that the matrix $\bar{A}(1)$ is substochastic, i. e., $\bar{A}(1)\mathbf{1}_m < \mathbf{1}_m$.

First introduce a new variable μ by $\mu = \frac{\lambda}{\rho}$. Then

$$P(\lambda) = P(\mu\rho) = \rho\big(\mu I - \rho^{-1} A(\mu\rho)\big).$$

Introduce matrix polynomials $\tilde{A}$ and $\tilde{P}$ via

$$\tilde{A}(\mu) = \rho^{-1} A(\mu\rho) \quad \text{and} \quad \tilde{P}(\mu) = \mu I - \tilde{A}(\mu), \tag{5-84}$$

i. e., $\tilde{P}(\mu) = \rho^{-1} P(\mu\rho)$ and $\tilde{A}_j = \rho^{j-1} A_j$ for $j \in \langle l \rangle_0$.

Then

$$\operatorname{spr} \tilde{A}(1) = \operatorname{spr}\left(\rho^{-1} A(\rho)\right) = \rho^{-1} \operatorname{spr}\left(A(\rho)\right) < \rho^{-1}\rho = 1.$$

Since $A(\rho)$ has a strictly positive eigenvector $v = (v_j)_{j=1}^n$ to the eigenvalue $\operatorname{spr}(A(\rho))$, the matrix $\tilde{A}(1) = \rho^{-1} A(\rho)$ has the same strictly positive eigenvector to the eigenvalue $\theta = \operatorname{spr}(\tilde{A}(1))$. Hence, the matrix D, defined by $D = \operatorname{diag}(v_1, \ldots, v_n)$ is nonsingular. Setting

$$\bar{A}(\mu) = D^{-1} \tilde{A}(\mu) D$$

one obtains

$$\bar{A}(1)\mathbf{1} = D^{-1}\tilde{A}(1)D\mathbf{1} = D^{-1}\tilde{A}(1)v = D^{-1}\theta v = \theta\mathbf{1} < \mathbf{1}.$$

Introduce the matrix polynomial $\bar{P}$ defined by

$$\bar{P}(\mu) = D^{-1}\tilde{P}(\mu)D = \mu I - \bar{A}(\mu) = \mu I - \sum_{j=0}^{l} \mu^j \bar{A}_j, \tag{5-85}$$

where $\bar{A}_j = \rho^{j-1} D^{-1} A_j D$ for $j \in \langle l \rangle_0$.

Then $\bar{P}$ is a 1–monic PFF with $\bar{A}(1)$ substochastic.

The next Proposition gives the connection between spectral factorizations of P and $\bar{P}$. Although we would like the left factor B to have the form given in (5-79), at this point it is more convenient to use $B(\lambda)B_0 = \sum_{j=0}^{l-1} \lambda^j B_j$, since this simplifies the following calculations and the desired algorithm. We will retrieve the desired form of B right at the end of that algorithm.

Proposition 5.3. *Let P with $P(\lambda) = \lambda I_n - A(\lambda)$ be a 1–monic PFP of degree l and suppose that $A(\rho)$ has a strictly positive eigenvector $v = (v_j)_{j=1}^n$ to the eigenvalue $\operatorname{spr} A(\rho)$. Suppose that P has a spectral right PFF with respect to $\mathbb{T}_\rho$, i. e.,*

$$P(\lambda) = B(\lambda)C(\lambda),$$

such that

$$\sigma(B) = \sigma(P) \cap (\mathbb{C} \setminus \bar{\mathbb{D}}), \quad \sigma(C) = \sigma(P) \cap \mathbb{D}$$

Then $\bar{P}$ has a spectral factorization with respect to the unit circle $\mathbb{T}$

$$\bar{P}(\mu) = \bar{B}(\mu)\bar{C}(\mu) \tag{5-86}$$

with

$$\bar{B}(\mu) = \sum_{j=0}^{l-1} \mu^j \bar{B}_j, \quad \bar{C}(\mu) = \mu I - \bar{C}_0$$

and

$$\sigma(\bar{B}) = \sigma(\bar{P}) \cap (\mathbb{C} \setminus \bar{\mathbb{D}}), \quad \sigma(\bar{C}) = \sigma(\bar{P}) \cap \mathbb{D}.$$

With $D = \operatorname{diag}(v_1, \ldots, v_n)$*, the factors* $B(\lambda)$ *and* $C(\lambda)$ *are obtained from* $\bar{B}$ *and* $\bar{C}$ *via*

$$B(\lambda) = \sum_{j=0}^{l-1} \lambda^j B_j, \quad \text{where } B_j = \rho^{-j} D \bar{B}_j D^{-1}, \tag{5-87}$$

$$C(\lambda) = \lambda I - C_0, \quad \text{where } C_0 = \rho D \bar{C}_0 D^{-1} \tag{5-88}$$

$$\tag{5-89}$$

Proof. If $v > 0$ is such that $A(\rho)v = \operatorname{spr} A(\rho)v$, then from (5-84) it follows that

$$\rho\tilde{A}(1)v = \operatorname{spr}\left(\rho\tilde{A}(1)\right)v = \rho \operatorname{spr}\left(\tilde{A}(1)\right)v,$$

i. e., v is a positive eigenvector of $\tilde{A}(1)$ corresponding to the eigenvalue $\operatorname{spr}\tilde{A}(1)$. (5-84) and (5-85) then imply that $\bar{P}(\mu) = \rho^{-1}D^{-1}P(\mu\rho)D$. Using the factorization for P we obtain

$$\bar{P}(\mu) = \rho^{-1}D^{-1}B(\mu\rho)C(\mu\rho)D = \rho^{-1}D^{-1}B(\mu\rho)DD^{-1}C(\mu\rho)D.$$

Define

$$\bar{B}(\mu) = D^{-1}B(\mu\rho)D \quad \text{and} \quad \bar{C}(\mu) = \rho^{-1}D^{-1}C(\mu\rho)D,$$

thus $\bar{P}(\mu) = \bar{B}(\mu)\bar{C}(\mu)$. Obviously λ is an eigenvalue of B (or C, respectively) if and only if λ/ρ is an eigenvalue of $\bar{B}$ (or $\bar{C}$, respectively). We have

$$\bar{B}(\mu) = D^{-1}\left(\sum_{j=0}^{l-1} \mu^j\rho^j B_j\right) D = \sum_{j=1}^{l-1} \mu^j \left(\rho^j D^{-1} B_j D\right) \quad \text{and}$$

$$\bar{C}(\mu) = \rho^{-1}D^{-1}\left(\mu\rho I - C_0\right)D = \mu I - \rho^{-1}D^{-1}C_0 D.$$

Setting $\bar{B}_j = \rho^j D^{-1} B_j D$ for $j \in \langle l-1 \rangle_0$ and $\bar{C}_0 = \rho^{-1}D^{-1}C_0 D$ immediately implies (5-87) and (5-88). □

In view of Proposition 5.3 we will now assume that $P(\lambda) = \lambda I - A(\lambda)$ is such that $A(1)$ is substochastic.

Now it will pay off that we started with a 1–monic PFP, or in other words, that we first passed over to the canonical reduction of P, since we can apply the following theorem, which can be found in [BLM05, Theorem 3.18] or [Mar88, Lemma 22.9]. We already made use of it in Section 4.1. Here, it is adapted to our current situation. We will write $P(X) = X - A(X) = X - \sum_{j=0}^{l} A_j X^j$.

Theorem 5.4. *Suppose that the 1–monic PFP* $P(\lambda) = \lambda I - A(\lambda)$ *is of degree* l *and that* $A(1)$ *is substochastic. If* P *has a spectral factorization*

$$P(\lambda) = B(\lambda)B_0C(\lambda) = \left(I_n - \sum_{j=1}^{l-1} \lambda^j B_j\right) B_0(\lambda I - C_0)$$

with respect to the unit circle $\mathbb{T}$ *(i. e.,* C_0 *is the spectral right root of* P *w.r.t.* $\mathbb{T}$*), then,* C_0 *is the unique solution of the equation* $P(X) = 0$ *such that* $\operatorname{spr}(C_0) < 1$.

Note that the equation $P(X) = 0$, or $X = \sum_{j=0}^{l} A_j X^j$, is of the form (5-80). Therefore, the CR-Algorithm applied to this equation delivers an approximation for the right root C_0 of P (for the notion of a right root, see Section 4.1).

5.2 Algorithm

For the remainder of this chapter we suppose that $P(\lambda) = \lambda I - A(\lambda)$ is a 1–monic PFP and $A(1) = \sum_{j=0}^{l} A_j$ is substochastic.

In this section we present an algorithm to compute the coefficients of the spectral factorization of a given m–monic matrix polynomial. As mentioned in the previous section, the method of cyclic reduction relies on the invertibility of the matrices $I - U_{1,1}^{(1)}, I - U_{1,1}^{(2)}, \dots$ which arise after the permutation in each step. To verify this, it is convenient to shortly introduce the notion of communicating and final classes according to [BLM05].

Let $G_A = (V, E)$ be the associated directed graph of A. For $i, j \in V$ define the relation $\sim$ via $i \sim j$ iff there is a path in G_A leading from i to j and a path in G_A leading from j to i. We say i **communicates** with j. Note that if i and j communicate via the paths P_1 and P_2 then i communicates with each vertex on P_1 and P_2. Hence, with the convention that each vertex in V communicates with itself, communication is an equivalence relation. We call an equivalence class induced by the communicate relation a **communicating class**. For instance, if A is irreducible, then V itself is the only communicating class in G_A.
Although no vertex of a communicating class $U \subset V$ communicates with a vertex which does not belong to U, there can still be paths leading out of it. We call a communicating class U such that there is no path leaving U, a **final class**. For simplicity of speech, we also say that the matrix A has a (final) communicating class.

We now return to the question if we can guarantee that each step of the CR algorithm is well defined. From [BLM05, Theorem 7.7 and Theorem 7.8] it follows that for each step k, the matrix $I - U_{1,1}^{(k)}$ is invertible if the matrix $A(1) = \sum_{j=0}^{l} A_j$ has only one final class. This is of course true if we initially started with a 1–monic PFP such that $A(1)$ is irreducible, since in that case, there is only one communicating class. However, if the initial polynomial was m–monic with $m > 1$, we performed a canonical reduction beforehand and $A(1) \in \mathbb{R}^{mn,mn}$ looks as follows

$$A(1) = \begin{bmatrix} \sum\limits_{j=m-1}^{l+m-1} \tilde{A}_j & \tilde{A}_{m-2} & \cdots & \tilde{A}_0 \\ I & & & \\ & \ddots & & \\ & & I & 0 \end{bmatrix}, \tag{5-90}$$

where $\tilde{A}_0, \tilde{A}_1, \dots, \tilde{A}_{l+m-1} \in \mathbb{R}^{n,n}$ are such that $\sum_{j=0}^{l+m-1} \tilde{A}_j$ is irreducible. Therefore, it remains to prove the following lemma.

Lemma 5.5. *The matrix $A(1)$ in (5-90) has exactly one final class.*

Proof. Consider the associated directed graph $G_{A(1)}$ of $A(1)$ and Figure 5 in Section 5.1. In that section, we already saw that since by assumption $\sum_{j=0}^{l} \tilde{A}_j$ is an irreducible matrix, we have that from each of the vertices in the first row of the graph there is a path to each column of the graph. More precisely, for each $i \in \langle n \rangle$ and each $j \in \langle n \rangle$ there is a path in $G_{A(1)}$ from i to $pn + j$ for some $p \in \langle m-1 \rangle_0$.

Now, in each column of $G_{A(1)} = (V, E)$, delete all vertices which are below the last vertex that has incoming edges from the first n vertices. In other words, for each $j \in \langle n \rangle$ do the following. Let $p_j \in \langle m-1 \rangle_0$ be the largest integer such that $(i, p_j n + j) \in E$ for some $i \in \langle n \rangle$. Delete each vertex $pn + j$ with $p_j < p \leqslant m - 1$ as well as its unique single outgoing edge. Note that the vertices $1, \dots, n$ are never deleted in this manner, since each column of the graph has at least one incoming edge from the vertices $1, \dots, n$. Completely analogously as in the proof of Proposition 5.1 it follows that the remaining set of vertices is a communicating class. It is also a final class, since in the process we only deleted vertices that had no incoming edges. Each of the vertices which where deleted in the process constitutes a communicating class only consisting of itself. Clearly none of them is a final class, since they all have an outgoing edge. Hence, $G_{A(1)}$ has precisely one final class. □

Now that we know that the method is applicable, we write it down explicitly step by step. For clarity, we will first summarize the strategy, since it also involves results from previous chapters.

Start with an m–monic PFP $P(\lambda) = \lambda^m I_n - A(\lambda)$ of degree l and a $\rho > 0$ such that $\phi_A(\rho) = \operatorname{spr} A(\rho) < \rho^m$. From Theorem 4.1 it follows that P has a spectral right PFF $C(\lambda) = \lambda^m I_n - \sum_{j=0}^{m-1} \lambda^j C_j$ with respect to the circle $\mathbb{T}_\rho$. Suppose furthermore that A is irreducible. Calculate the canonical reduction $\mathcal{P}(\lambda) = \lambda I_{mn} - \mathcal{A}(\lambda)$, see Corollary 3.3, which is a 1–monic PFF of degree $l-m+1$ which satisfies $\phi_{\mathcal{A}}(\rho) < \rho$. Perform a scaling of $\mathcal{P}$ to a polynomial $\bar{\mathcal{P}}$, $\bar{\mathcal{P}}(\lambda) = \lambda I_{mn} - \bar{\mathcal{A}}(\lambda)$ such that $\bar{\mathcal{A}}(1)$ is substochastic, according to Section 5.1. $\bar{\mathcal{A}}$ satisfies $\phi_{\bar{\mathcal{A}}}(1) < 1$, hence, it has a spectral right root $\bar{\mathcal{C}}_0$ w.r.t. the unit circle. Apply the CR-algorithm as given in [BLM05, Section 7.4] to obtain an approximation for the minimal solution of the equation $\bar{\mathcal{P}}(\mathcal{X}) = 0$ or $\mathcal{X} = \sum_{j=0}^{l-m+1} \bar{\mathcal{A}}_j \mathcal{X}^j$, which coincides with $\bar{\mathcal{C}}_0$, due to Theorem 5.4. Use Lemma 4.2 to obtain the coefficients $\bar{\mathcal{B}}_0, \dots, \bar{\mathcal{B}}_{l-m}$ of the left factor of $\bar{\mathcal{P}}$ and then Proposition 5.3 to calculate the coefficients $\mathcal{B}_0, \dots, \mathcal{B}_{l-m}, \mathcal{C}_0$ of the factorization of $\mathcal{P}$. From Corollary 3.6 and Remark 3.5 obtain the coefficients $B_0, B_1, \dots, B_{l-m}, C_0, \dots, C_{m-1}$ of the desired spectral factorization of the original PFP P. B_0 is a nonsingular M-matrix, see Proposition 4.7.

Algorithm 5.6 (Spectral factorization for $P(\lambda) = \lambda I_n - A(\lambda) = \lambda^m I_n - \sum_{j=0}^{l} \lambda^j A_j$).

INPUT: The coefficients $A_0, \dots, A_l$ and $\rho > 0$ such that $\operatorname{spr} A(\rho) < \rho^m$.

OUTPUT: The coefficients $B_0, B_1, \dots, B_{l-m}, C_0, \dots, C_{m-1}$ of the factorization

$$P(\lambda) = \Big(I_n - \sum_{j=1}^{l-m} \lambda^j B_j\Big) B_0 \Big(\lambda^m I_n - \sum_{j=0}^{m-1} \lambda^j C_j\Big).$$

COMPUTATION:

1. Build the matrices
$$\mathcal{A}_0 = \begin{bmatrix} A_{m-1} & \cdots & \cdots & A_0 \\ I & & & 0 \\ & \ddots & & \vdots \\ & & I & 0 \end{bmatrix} \quad \text{and} \quad \mathcal{A}_j = \begin{bmatrix} A_{j+m-1} & & & \\ & 0 & & \\ & & \ddots & \\ & & & 0 \end{bmatrix} \quad (j \in \langle l-m+1 \rangle)$$
and set $\mathcal{A}(\rho) = \sum_{j=0}^{l-m+1} \rho^j \mathcal{A}_j$.
2. Compute a positive eigenvector v of $\rho^{-1}\mathcal{A}(\rho)$ to the eigenvalue $\rho^{-1} \operatorname{spr} \mathcal{A}(\rho)$.

3. Set $\mathcal{D} = \operatorname{diag} v$ and $\bar{\mathcal{A}}_j = \rho^{j-1}\mathcal{D}^{-1}\mathcal{A}_j\mathcal{D}$ for $j \in \langle l \rangle_0$.

4. Apply the CR algorithm [BLM05, Section 7.4] to compute the minimal solution $\bar{\mathcal{C}}_0$ of the matrix equation

$$\mathcal{X} = \sum_{j=0}^{l-m+1} \bar{\mathcal{A}}_j \mathcal{X}^j.$$

5. Set

$$\bar{\mathcal{B}}_k = \delta_{0k} I_{mn} - \sum_{j=0}^{l-m-k} \bar{\mathcal{A}}_{k+j+1} \bar{\mathcal{C}}_0^j \qquad (k \in \langle l-m \rangle_0).$$

6. Set

$$\mathcal{C}_0 = \rho \mathcal{D} \bar{\mathcal{C}}_0 \mathcal{D}^{-1} \quad \text{and} \quad \mathcal{B}_k = \rho^{-k} \mathcal{D} \bar{\mathcal{B}}_k \mathcal{D}^{-1} \qquad (k \in \langle l-m \rangle_0).$$

7. Set

$$C_k = (\mathcal{C}_0)_{1(m-k)} \qquad (j \in \langle m-1 \rangle_0),$$

$$B_0 = (\mathcal{B}_0)_{11} \quad \text{and} \quad B_k = -(\mathcal{B}_k)_{11} B_0^{-1} \qquad (k \in \langle l-m \rangle).$$

6 Conclusions

In this thesis we study spectral properties of analytic m–monic operator- and matrix functions $F(\lambda) = \lambda^m I - A(\lambda)$. A major focus lies on m–monic operator- and matrix polynomials with coefficients that satisfy certain nonnegativity conditions.

Chapter 1 establishes a general framework for the investigation of spectral properties of m–monic Banach algebra functions. A major role is played by a geometrically convex scalar function ϕ_A associated with the spectral radius of $A(\lambda)$. This function was introduced in [FN05a].

Eigenvalues of m–monic operator functions which lie on some circles around zero are rotation invariant with respect to angles corresponding to certain roots of unity if the coefficients of the function are self-adjoint and satisfy some condition which is closely related to a condition on the function ϕ_A at some point $\rho > 0$. This is investigated in Chapter 2 and extends results of [Wim08] and [SW10].

Very crucial for the investigation of spectral properties of m–monic operator- and matrix polynomials is the canonical reduction from the m–monic to the 1–monic case. It is a generalization of the linearization via the first companion form and allows the application of fixpoint iterations, which are an suitable tool for the study of factorizations of 1–monic operator- and matrix polynomials. A one–to–one correspondence between factorizations of the m–monic polynomial and its canonical reduction guarantees that the original polynomial has a factorization if and only if its canonical reduction has one. Formulas which express one factorization in terms of the other one are available. Furthermore, in the case of matrix polynomials, it is possible to calculate Jordan chains of the original polynomial from the Jordan chains of its canonical reduction and vice versa. The canonical reduction and its properties are established in Chapter 3.

Spectral properties and factorization results of m–monic matrix polynomials with entrywise nonnegative coefficients are presented in Chapter 4.
Using fixpoint iterations, a condition for the existence of spectral factorizations of an 1–monic matrix polynomial P with nonnegative coefficients is given. Those factorizations separate the eigenvalues of P with respect to certain circles. Via canonical reduction, this result can be transferred to the m–monic case. This is a special case of an abstract result in ordered Banach algebras given in [FN05b] and its proof uses only matrix theoretical concepts. This is given in Sections 4.1 and 4.2.
m–monic matrix polynomials P with entrywise nonnegative coefficients such that the sum of the coefficients is irreducible can have eigenvalues that have a certain rotation invariance. This symmetry is very similar to the rotation invariance of peripheral eigenvalues of entrywise nonnegative irreducible matrices which is a consequence of the well known Perron–Frobenius theory. The study of an infinite graph associated with P is key for the analysis.
A description of spectral factorizations, their existence and their eigenvalues in the irreducible case is given via the characterization of eight cases which can occur. These are extensions to results of [FN05a] and are studied in Section 4.3.

Assumed that an P satisfies the condition for the existence of a spectral factorization (Section 4.2), a numerical algorithm for the calculation of the corresponding factors is available. Its core is constituted by a version of a cyclic reduction method suited for a certain type of Markov chains which is presented in [BLM05]. A suitable transformation of the factorization problem to a Markov problem allows its application. This is done in Chapter 5.

References

[AHKM03] T. Y. Azizov, V. Hardt, N. D. Kopachevsky, and R. Mennicken. On the problem of small motions and normal oscillations of a viscous fluid in a partially filled container. *Math. Nachr.*, 248/249:3–39, 2003.

[AKL68] N. K. Askerov, S. G. Kreĭn, and G. I. Laptev. The problem of the oscillations of a viscous liquid and the operator equations connected with it. *Funkcional. Anal. i Priložen.*, 2(2):21–31, 1968.

[AKS97] T. Ya. Azizov, N. D. Kopachevsky, and L. I. Sukhocheva. On eigenvalues of selfadjoint pencils with parameter. In *Operator theory, operator algebras and related topics (Timişoara, 1996)*, pages 37–50. Theta Found., Bucharest, 1997.

[AP97] P. Amodio and M. Paprzycki. A cyclic reduction approach to the numerical solution of boundary value ODEs. *SIAM J. Sci. Comput.*, 18(1):56–68, 1997.

[Aup91] B. Aupetit. *A Primer on Spectral Theory.* Universitext. Springer-Verlag, New York, 1991.

[AV04] E. N. Antoniou and S. Vologiannidis. A new family of companion forms of polynomial matrices. *Electron. J. Linear Algebra*, 11:78–87 (electronic), 2004.

[Bar83] S. Barnett. *Polynomials and Linear Control Systems*, volume 77 of *Monographs and Textbooks in Pure and Applied Mathematics.* Marcel Dekker Inc., New York, 1983.

[Bau85] H. Baumgärtel. *Analytic Perturbation Theory for Matrices and Operators*, volume 15 of *Operator Theory: Advances and Applications.* Birkhäuser Verlag, Basel, 1985.

[BD71] F. F. Bonsall and J. Duncan. *Numerical Ranges of Operators on Normed Spaces and of Elements of Normed Algebras*, volume 2 of *London Mathematical Society Lecture Note Series.* Cambridge University Press, London, 1971.

[BD73] F. F. Bonsall and J. Duncan. *Numerical Ranges. II.* Cambridge University Press, New York, 1973. London Mathematical Society Lecture Notes Series, No. 10.

[BG94] S. Bondeli and W. Gander. Cyclic reduction for special tridiagonal systems. *SIAM J. Matrix Anal. Appl.*, 15:321–330, 1994.

[BGN70] B. L. Buzbee, G. H. Golub, and C. W. Nielson. On direct methods for solving poisson's equations. *SIAM J. Numer. Anal.*, 7(4):627–656, 1970.

[BLM05] D. A. Bini, G. Latouche, and B. Meini. *Numerical Methods for Structured Markov Chains*. Numerical Mathematics and Scientific Computation. Oxford University Press, New York, 2005. Oxford Science Publications.

[BM09] D. A. Bini and B. Meini. The cyclic reduction algorithm: from Poisson equation to stochastic processes and beyond. *Numer. Alg.*, 51(1):23–60, 2009.

[Bon55] F. F. Bonsall. Endomorphisms of a partially ordered vector space without order unit. *J. London Math. Soc.*, 30:144–153, 1955.

[BP94] A. Berman and R. J. Plemmons. *Nonnegative Matrices in the Mathematical Sciences*, volume 9 of *Classics in Applied Mathematics*. Society for Industrial and Applied Mathematics (SIAM), Philadelphia, PA, 1994. Revised reprint of the 1979 original.

[BS99] A. Böttcher and B. Silbermann. *Introduction to Large Truncated Toeplitz Matrices*. Universitext. Springer-Verlag, New York, 1999.

[Con85] J. B. Conway. *A Course in Functional Analysis*. Springer-Verlag, New York, 1985.

[FN91] K.-H. Förster and B. Nagy. Some properties of the spectral radius of a monic operator polynomial with nonnegative compact coefficients. *Integral Equations Operator Theory*, 14(6):794–805, 1991.

[FN05a] K.-H. Förster and B. Nagy. On nonmonic quadratic matrix polynomials with nonnegative coefficients. In *Operator theory in Krein spaces and nonlinear eigenvalue problems*, volume 162 of *Oper. Theory Adv. Appl.*, pages 145–163. Birkhäuser, Basel, 2005.

[FN05b] K.-H. Förster and B. Nagy. Spectral properties of operator polynomials with nonnegative coefficients. In *Operator theory and indefinite inner product spaces*, volume 163 of *Oper. Theory Adv. Appl.*, pages 147–162. Birkhäuser, Basel, 2005.

[GHT96] H. R. Gail, S. L. Hantler, and B. A. Taylor. Spectral analysis of $M/G/1$ and $G/M/1$ type Markov chains. *Adv. in Appl. Probab.*, 28(1):114–165, 1996.

[GHT98] H. R. Gail, S. L. Hantler, and B. A. Taylor. Matrix-geometric invariant measures for $G/M/1$ type Markov chains. *Comm. Statist. Stochastic Models*, 14(3):537–569, 1998.

[GKL88] I. Gohberg, M. A. Kaashoek, and P. Lancaster. General theory of regular matrix polynomials and band Toeplitz operators. *Integral Equations Operator Theory*, 11(6):776–882, 1988.

[GKS03] I. Gohberg, M. A. Kaashoek, and I. M. Spitkovsky. An overview of matrix factorization theory and operator applications. In *Factorization and integrable systems (Faro, 2000)*, volume 141 of *Oper. Theory Adv. Appl.*, pages 1–102. Birkhäuser, Basel, 2003.

[GLR82] I. Gohberg, P. Lancaster, and L. Rodman. *Matrix Polynomials*. Academic Press Inc. [Harcourt Brace Jovanovich Publishers], New York, 1982. Computer Science and Applied Mathematics.

[Hel76] D. Heller. Some aspects of the cyclic reduction algorithm for block tridiagonal linear systems. *SIAM J. Numer. Anal.*, 13(4):484–496, 1976.

[Heu06] H. Heuser. *Funktionalanalysis*. Teubner Verlag, Stuttgart, 2006. 4. überarbeitete Auflage.

[HJ85] R. A. Horn and C. R. Johnson. *Matrix Analysis*. Cambridge University Press, Cambridge, 1985.

[HJ91] R. A. Horn and C. R. Johnson. *Topics in Matrix Analysis*. Cambridge University Press, Cambridge, 1991.

[HK76] W. K. Hayman and P. B. Kennedy. *Subharmonic Functions. Vol. I.* Academic Press [Harcourt Brace Jovanovich Publishers], London, 1976. London Mathematical Society Monographs, No. 9.

[Kat76] T. Kato. *Perturbation Theory for Linear Operators*. Classics in Mathematics. Springer-Verlag, Berlin, 1976.

[KL68] S. G. Kreĭn and G. I. Laptev. On the problem of the motion of a viscous fluid in an open vessel. *Funkcional. Anal. i Priložen.*, 2(1):40–50, 1968.

[KLS89] M. A. Krasnosel′skij, J. A. Lifshits, and A. V. Sobolev. *Positive linear systems*, volume 5 of *Sigma Series in Applied Mathematics*. Heldermann Verlag, Berlin, 1989. The method of positive operators, Translated from the Russian by Jürgen Appell.

[Kre89] E. Kreyszig. *Introductory Functional Analysis with Applications*. Wiley Classics Library. John Wiley & Sons Inc., New York, 1989.

[Kuc85] M. Kuczma. *An introduction to the theory of functional equations and inequalities*, volume 489 of *Prace Naukowe Uniwersytetu Śląskiego w Katowicach [Scientific Publications of the University of Silesia]*. Uniwersytet Śląski, Katowice, 1985. Cauchy's equation and Jensen's inequality, With a Polish summary.

[LP05] P. Lancaster and P. Psarrakos. A note on weak and strong linearizations of regular matrix polynomials. *Numerical Analysis Report No.470, Manchester Centre for Computational Mathematics, Manchester, England*, 2005.

[LR99] G. Latouche and V. Ramaswami. *Introduction to Matrix Analytic Methods in Stochastic Modelling*, volume 5 of *ASA-SIAM Series on Statistics and Applied Probability*. SIAM, Philadelphia, PA, 1999.

[Mar88] A. S. Markus. *Introduction to the Spectral Theory of Polynomial Operator Pencils*, volume 71 of *Translations of Mathematical Monographs*. American Mathematical Society, Providence, RI, 1988. Translated from the Russian by H. H. McFaden, Translation edited by Ben Silver, With an appendix by M. V. Keldysh.

[Mei06] B. Meini. Nonlinear matrix equations and structured linear algebra. *Linear Algebra Appl.*, 413:440–457, 2006.

[Min88] H. Minc. *Nonnegative Matrices.* Wiley-Interscience Series in Discrete Mathematics and Optimization. John Wiley & Sons Inc., New York, 1988. A Wiley-Interscience Publication.

[MMMM06] D. S. Mackey, N. Mackey, C. Mehl, and V. Mehrmann. Vector spaces of linearizations for matrix polynomials. *SIAM J. Matrix Anal. Appl.*, 28(4):971–1004 (electronic), 2006.

[Mor69] L. J. Mordell. *Diophantine Equations.* Pure and Applied Mathematics, Vol. 30. Academic Press, London, 1969.

[Nag07] B. Nagy. On the degree reduction of matrix polynomials. *Private communication*, 2007.

[Neu89] M. F. Neuts. *Structured stochastic matrices of $M/G/1$ type and their applications*, volume 5 of *Probability: Pure and Applied.* Marcel Dekker Inc., New York, 1989.

[Neu94] M. F. Neuts. *Matrix-Geometric Solutions in Stochastic Models.* Dover Publications Inc., New York, 1994. An algorithmic approach, Corrected reprint of the 1981 original.

[PT04] P. J. Psarrakos and M. J. Tsatsomeros. A primer of Perron-Frobenius theory for matrix polynomials. *Linear Algebra Appl.*, 393:333–351, 2004.

[Rau92] R. T. Rau. On the peripheral spectrum of monic operator polynomials with positive coefficients. *Integral Equations Operator Theory*, 15(3):479–495, 1992.

[Rod89] L. Rodman. *An Introduction to Operator Polynomials*, volume 38 of *Operator Theory: Advances and Applications.* Birkhäuser Verlag, Basel, 1989.

[RR96] H. Rautenheimer and S. Rode. Cones in Banach algebras. *Indag. Math. (N.S.)*, 7(4):489–502, 1996.

[Sch71] H. H. Schaefer. *Topological Vector Spaces.* Springer-Verlag, New York, 1971. Third printing corrected, Graduate Texts in Mathematics, Vol. 3.

[Sch74] H. H. Schaefer. *Banach Lattices and Positive Operators.* Springer-Verlag, New York, 1974. Die Grundlehren der mathematischen Wissenschaften, Band 215.

[Sch86] H. Schneider. The influence of the marked reduced graph of a nonnegative matrix on the Jordan form and on related properties: a survey. In *Proceedings of the symposium on operator theory (Athens, 1985)*, volume 84, pages 161–189, 1986.

[Sin71] A. M. Sinclair. The norm of a Hermitian element in a Banach algebra. *Proc. Amer. Math. Soc.*, 28(2):446–450, 1971.

[Ste05] J. Steuding. *Diophantine Analysis.* Discrete Mathematics and its Applications (Boca Raton). Chapman & Hall/CRC, Boca Raton, FL, 2005.

[Suk97] L. I. Sukhochëva. On some spectral properties of a quadratic selfadjoint matrix pencil with dominating main diagonals. *Mat. Zametki*, 61(3):381–390, 1997.

[SW10] J. Swoboda and H. K. Wimmer. Spectraloid operator polynomials, the approximate numerical range and an Eneström-Kakeya theorem in Hilbert space. *Studia Math.*, 198(3):279–300, 2010.

[Swe88] R. A. Sweet. A parallel and vector variant of the cyclic reduction algorithm. *SIAM J. Sci. Stat. Comp.*, 9(4):761–765, 1988.

[Wer00] D. Werner. *Funktionalanalysis.* Springer-Verlag, Berlin, extended edition, 2000.

[Wim08] H. K. Wimmer. Polynomial matrices with Hermitian coefficients and a generalization of the Eneström-Kakeya theorem. *Oper. Matrices*, 2(3):443–454, 2008.

[Żel73] W. Żelazko. *Banach algebras.* Elsevier Publishing Co., Amsterdam-London-New York, 1973. Translated from the Polish by M. E. Kuczma.

MIX
Papier aus verantwortungsvollen Quellen
Paper from responsible sources
FSC® C105338

Printed by Books on Demand GmbH, Norderstedt / Germany